Consumption Choices

DEVELOPMENT ECONOMICS AND POLICY

Series edited by Franz Heidhues †, Joachim von Braun,
Ulrike Grote and Manfred Zeller

Vol. 80

Consumption Choices
The effects of food production, markets and preferences on diets in India

Till Ludwig

Bibliographic Information published by the Deutsche Nationalbibliothek
The Deutsche Nationalbibliothek lists this publication in the Deutsche Nationalbibliografie; detailed bibliographic data is available in the internet at http://dnb.d-nb.de.

Library of Congress Cataloging-in-Publication Data
A CIP catalog record for this book has been applied for at the Library of Congress.

Zugl.: Bonn, Univ., Diss., 2019

Printed by CPI books GmbH, Leck

D5
ISSN 0948-1338
ISBN 978-3-631-79735-8 (Print)
E-ISBN 978-3-631-80089-8 (E-Book)
E-ISBN 978-3-631-80090-4 (EPUB)
E-ISBN 978-3-631-80091-1 (MOBI)
DOI 10.3726/b16086

© Peter Lang GmbH
Internationaler Verlag der Wissenschaften
Berlin 2019
All rights reserved.

Peter Lang – Berlin · Bern · Bruxelles · New York ·
Oxford · Warszawa · Wien

All parts of this publication are protected by copyright. Any utilisation outside the strict limits of the copyright law, without the permission of the publisher, is forbidden and liable to prosecution. This applies in particular to reproductions, translations, microfilming, and storage and processing in electronic retrieval systems.

This publication has been peer reviewed.

www.peterlang.com

Summary

The choice of which food to consume is often an individual choice. Yet, food and nutrition insecurity presents a situation that those who suffer from it have rarely chosen consciously. Availability of foods, their accessibility and food of good quality are necessary conditions to food security. A lack of these or unhealthy food preferences of the consumers are factors contributing to poor diets and malnutrition. High prevalence of food and nutrition insecurity often occurs in rural areas. These areas are lacking economic opportunities and malnutrition often coincides with poverty. Monotonous diets consisting of few food varieties are the norm and one reason for poor nutrition outcomes. At the same time, many malnourished families are food-producing farmers leading to a puzzle of cause and effect of malnutrition.

The present research aims to analyze the drivers of dietary diversity of individuals and households. The study areas are rural regions in India whose population has a high rate of malnutrition and which is prone to various risks. The results indicate that tangible factors such as food production and market access, but also intangible factors such as economic preferences play a vital role in achieving and maintaining a diverse and secure nutrition.

Agricultural production is the starting point for most food value chains. Clearly, diet choices can only be made on the basis of available foods. It is not so clear if diverse diets are a result of diverse agricultural production or if other factors such as markets are mediating these. The first research objective examines the link between production diversity and dietary diversity of smallholder farmers. We identify that a diverse production does affect diverse food consumption; if production diversity is increases by 1 food group, the dietary diversity of women increases up to 18.8%. However, we also find that market access has a much stronger effect and can even negate direct effects of production diversity. Markets can provide foods that are not produced by smallholder farmers and further increase the accessible food choices.

Given a certain food variety and availability, individual preferences still steer actual food consumption. Preferences are core drivers of diets; taste preferences guide us to prefer one food over another. But there are also deep preferences that guide our behavior subconsciously. Risk aversion or risk affinity, altruism or egoistic behavior are character traits that affect daily choices. The second research objective provides the theoretical foundation on how these preferences affect food consumption behavior. On the basis of the expected utility theory, we develop a model that predicts the effects. The third research objective tests the model empirically. We utilize an innovative survey methodology to elicit the preferences in rural areas of India. We show that risk preference and altruism do influence dietary choices and, thus, nutrition security. An increase of 10 percentage points in risk taking increases the dietary diversity score by up to 1.4%; altruistic behavior of the household head improves the nutrition by up to 3.0%.

Zusammenfassung

Die Wahl, welche Lebensmittel konsumiert werden sollen, ist oft eine individuelle Entscheidung. Ernährungsunsicherheit beschreibt jedoch eine Situation, die unternährte Menschen selten bewusst gewählt haben. Die Verfügbarkeit von Lebensmitteln, ihre Zugänglichkeit und Qualität sind notwendige Bedingungen für die Ernährungssicherung. Das Fehlen dieser oder unterschiedliche Präferenzen des Verbrauchers sind Faktoren, die zu einer schlechten Ernährung und Unterernährung beitragen. Insbesondere im ländlichen Raum finden sich hohe Raten der Ernährungsunsicherheit. Den ländlichen Gebieten mangelt es häufig an wirtschaftlichen Möglichkeiten und Mangelernährung geht meist mit Armut einher. Monotone Diäten, die aus wenigen Nahrungsmittelsorten bestehen, sind die Norm und ein Grund für die körperliche Unterentwicklung. Die Frage von Ursache und Wirkung von Mangelernährung wird insbesondere dadurch aufgeworfen, dass viele mangelernährte Familien Landwirte sind, die Nahrungsmittel selbst produzieren.

Vor diesem Hintergrund zielt die vorliegende Forschung darauf ab, die Determinanten einer ausgewogenen Ernährung zu analysieren. Das Untersuchungsgebiet umfasst den ländlichen Raum Indiens, dessen Bevölkerung eine hohe Prävelenz an Mangelernährung aufweist und die vielfältigen Risiken ausgesetzt ist. Die Ergebnisse zeigen, dass greifbare Faktoren wie die Nahrungsmittelproduktion und der Marktzugang, aber auch immaterielle Faktoren wie individuelle Präferenzen eine entscheidende Rolle bei der Erreichung und Aufrechterhaltung einer vielfältigen und sicheren Ernährung spielen.

Die landwirtschaftliche Produktion ist der Ausgangspunkt der meisten Nahrungsmittelwertschöpfungsketten. Die produzierten Lebensmittel bilden die Grundlage jeglicher Konsumentscheidung. Es ist nicht eindeutig, ob eine abwechslungsreiche Ernährung auf eine vielfältige landwirtschaftliche Produktion zurückzuführen ist oder ob andere Faktoren wie zum Beispiel Märkte die Ernährung beeinflussen. Daher untersucht das erste Forschungsziel den Zusammenhang zwischen Produktionsvielfalt und Ernährungsvielfalt der Kleinbauern. Wir stellen fest, dass eine vielfältige Produktion einen diversen Lebensmittelkonsum beeinflusst. Eine Erhöhung der Produktionsdiversität um eine Einheit, erhöht die Nahrungsmitteldiversität bei Frauen um bis zu 18,8%. Wir stellen allerdings auch fest, dass der Marktzugang einen viel stärkeren Effekt hat und dieser sogar den positiven Effekt der Produktionsdiversität negieren kann. Märkte können Lebensmittel bereitstellen, die nicht von Kleinbauern erzeugt werden, und die Auswahl der verfügbaren Lebensmittel weiter verbessern.

Bei einer bestimmten Lebensmittelvielfalt und -verfügbarkeit steuern die individuellen Vorlieben letztendlich den tatsächlichen Lebensmittelkonsum. Präferenzen sind die Haupttreiber von Diäten; sensorische Präferenzen leiten uns dazu, ein Nahrungsmittel einem anderen vorzuziehen. Es gibt aber auch „tiefe" Präferenzen, die unser Verhalten unbewusst leiten. Risikoaversion oder Risikoaffinität, Altruismus oder egoistisches Verhalten sind Charaktereigenschaften, die tägliche Entscheidungen beeinflussen. Das zweite Forschungsziel liefert die theoretische Grundlage dafür wie Präferenzen das Verbrauchsverhalten beeinflussen. Basierend auf der *expected utility theory* entwickeln wir ein Modell, das die Auswirkungen von Risikoaffinität und Altruismus auf die Ernährung bestimmt. Das dritte Forschungsziel testet dieses Modell empirisch. Dabei verwenden wir eine innovative Erhebungsmethode, um individuelle Präferenzen im ländlichen Raum Indiens zu ermitteln. Wir zeigen, dass Risikopräferenz und Altruismus die Ernährungsgewohnheiten und somit die Ernährungssicherheit beeinflussen. Eine um 10 Prozentpunkte höhere Risikobereitschaft erhöht die Nahrungsmitteldiversität um bis zu 1,4%; ein altruistischeres Verhalten des Familienvaters erhöht die Nahrungsmitteldiversität um bis zu 3,0%.

Acknowledgements

I owe my deepest gratitude and appreciation to many people across the globe. Foremost, I thank my supervisor Professor Joachim von Braun for enabling me to conduct this research, for guiding me and advising me, for supporting me during my research in Bonn and in India and for being always available through the past three years. Without his support, this thesis clearly would have not been possible. I gratefully acknowledge Professor Matin Qaim for agreeing to be my second supervisor and for his comments. I also thank Professor Thomas Heckelei for being part of my academic committee.

My sincere gratitude goes to Guruji Guido Lüchters, from whom I learned academic essentials but who, much more significantly, encouraged me and uplifted me. I would like to acknowledge the valuable feedback by Mekbib Haile, Jan Brockhaus and Professor Matthias Kalkuhl. I thank Professor Arijita Dutta of the University of Calcutta, Philippe Dresruesse, Anshuman Das and Sweta Banerjee of Welthungerhilfe, Babita Sinha of Pravah as well as Jonathan Ziebula and Rajeshwari S M of GIZ for their organizational support in India without whom the data collection would have not been possible. I received much support by student assistants in Germany and India for which I am thankful to Davide Pesenti, Claudia Witkowski, Jayashis Ghosh, Madhurima Saha and Gayatri Mitra. I also appreciate the support by John Stravellis for editing the thesis. At the fantastic institution of ZEF I thank Gisela Ritter-Pilger, Günther Manske and Maike Retat-Amin and many more who made the time very enjoyable and productive. I am thankful to Professor Elisabeth Sadoulet and Professor Alain de Janvry for their kind sponsoring of my stay at the University of California, Berkeley.

I thank my data enumeration teams in India who were essential to this research. Namely, in Jharkhand I thank Deepak Kumar Singh, Abhishek Burman, Budhan Kumar, Jitendra Kumar Singh, Kumod Ranjan, Pulak Kumar Raut, Rajkishor Murmu, Raju Mandal, Teklal Murmu and Vijay Kumar Yadav. In West Bengal I thank Santanu Dasgupta, Abhijit Dutta, Indrajit Roy, Rajat Adhikari, Somali Sanyal and Sudip Kumar Samanta. In Karnataka I thank Adarsh, Raghunath, Meghana, Nagesh, Kumbi Vishwapariya and Manjunath.

Every research projects needs donors. I thank the German Federal Ministry for Economic Cooperation and Development, the foundation fiat panis and the German Academic Exchange Service for their financial support.

I thank all friends and colleagues at ZEF and beyond who I could meet over the course of the research and without whom my time would not have been as delightful and stimulating as it has been. I appreciate the fine arts and those who showed and introduced me to these as well as the music that kept me going despite the many obstacles. Thanks for letting me take a glimpse at the true value of life.

Contents

Summary 5

Zusammenfassung 7

Acknowledgements 9

List of Appendices 15

List of Tables 17

List of Figures 19

Abbreviations 21

1 Introduction 23
   1.1 Background and Motivation 23
   1.2 Research Objective and Questions 27
   1.3 Conceptual Framework 29
   1.4 Nutrition – Engine of Development 35
   1.5 Measurement of Nutrition 38
   1.6 Research Design 42
      1.6.1 Study sites 42
      1.6.2 Sampling 45

2 Focus on India: Food Prices and Food Security ... 49
   2.1 Introduction 49
   2.2 Recent Economic Development 49
   2.3 Food Price Trends 51
   2.4 Food Consumption Trends 53

3 Production Diversity and Dietary Diversity ... 59
- 3.1 Introduction ... 59
- 3.2 Empirical Evidence and Research Questions ... 61
- 3.3 Theoretical Model ... 63
- 3.4 Data ... 69
  - 3.4.1 Dependent variables ... 70
  - 3.4.2 Main explanatory variables ... 75
- 3.5 Methods ... 77
  - 3.5.1 Estimation strategy ... 77
  - 3.5.2 Identification strategy ... 78
- 3.6 Results ... 81
  - 3.6.1 Primary results: Effects of production diversity ... 81
  - 3.6.2 Secondary results: Effects of markets ... 88
  - 3.6.3 Robustness checks ... 94
- 3.7 Discussion ... 96
- 3.8 Conclusion ... 98

4 Considering Preferences for Food Consumption ... 101
- 4.1 Introduction ... 101
- 4.2 Related Literature and Contribution ... 103
- 4.3 Theoretical Model ... 107
  - 4.3.1 Preference model for nutrition ... 109
  - 4.3.2 Discussion of the model ... 115
  - 4.3.3 Model extension for altruism ... 118
  - 4.3.4 Limitations of the model ... 120
- 4.4 Conclusion ... 122

5 Effects of Preferences on Food Consumption ... 125
- 5.1 Introduction ... 125
- 5.2 Distribution of Preferences ... 127
- 5.3 Theoretical Model ... 130
- 5.4 Data ... 132
  - 5.4.1 Preference elicitation ... 132
  - 5.4.2 Main variables ... 140
- 5.5 Estimation Strategy ... 140
- 5.6 Results ... 146

		5.6.1 Primary results	146
		5.6.2 Secondary results	151
		5.6.3 Robustness checks	157
	5.7	Conclusion	161

6 Concluding Remarks ... 163
   6.1 Summary and Contribution ... 163
   6.2 Policy Implications, Limitations and Further Research ... 165

Bibliography ... 235

List of Appendices

A.1 Prevalence of Micronutrient Deficiency 172
A.2 Micronutrients, Sources and Consequences of Their Lack in Nutrition 173
A.3 Food Groups of Food Intake Indicators 174
A.4 FIES Questions 176
A.5 Random Walk Sampling Technique 177
A.6 Sampling 178
B.1 Checks for Instrument Variable 182
B.2 Effects of Policies 184
B.3 Market Access Effects 185
B.4 Production Diversity and Income 191
C.1 Solving Integral for Mean Variance Form 193
C.2 Differentiation for Preference Model 195
C.3 Solving for Optimal Savings Rate 199
D.1 Spatial Distribution of Preferences in the World 202
D.2 Distribution of Preferences in the World 204
D.3 Preferences and Income in the World 206
D.4 Spatial Distribution of Preferences in India 208
D.5 Preferences and Income in India 211
D.6 Distribution of Preferences in India 214
D.7 Survey Preference Module 216
D.8 Calculation of Preferences 220
D.9 Distribution of Preferences 221
D.10 Summary Statistics 224
D.11 Robustness Checks 226

List of Tables

Table 1.1	Characteristics of rural study regions	44
Table 3.1	Description of main variables	71
Table 3.2	Summary statistics for variables	73
Table 3.3	Impact of production diversity on dietary diversity of women	82
Table 3.4	Impact of production diversity on minimum dietary diversity of women	89
Table 3.5	Impact of market access on dietary diversity of women	92
Table 3.6	Consumption of food groups by market visits	94
Table 5.1	Items of survey preference module	134
Table 5.2	Scale of preference traits	135
Table 5.3	Preference traits statistics	136
Table 5.4	Description of variables	141
Table 5.5	Summary statistics for variables	144
Table 5.6	Regression of spouse's risk levels and household head's altruism levels on dietary diversity	148
Table 5.7	Regression of spouse's risk levels and household head's altruism levels on dietary diversity, detailed	149
Table 5.8	Correlation of risk levels with food and nutrition security indicators	153
Table 5.9	Correlation of various preferences with nutrition intake variables	154
Table 5.10	Point elasticities of risk on income	155
Table 5.11	Decomposing risk preference	157
Table 5.12	Correlation of various groups	159

List of Figures

Figure 1.1	Prevalence of anemia among women (in percent)	24
Figure 1.2	UNICEF conceptual framework of the determinants of nutritional status	31
Figure 1.3	Conceptual framework	34
Figure 1.4	Criteria for risk of deficiency and excess food intake	36
Figure 1.5	The cycle of micronutrient inadequacies	37
Figure 2.1	Trends in economic growth, agricultural production and population growth in India, 1961-2016	50
Figure 2.2	Food price indices in India, three year averages from 1970 to 2017	52
Figure 2.3	Food supply in India, 1976-2013	53
Figure 2.4	Food supply of unhealthy foods in India, 1976-2013	54
Figure 2.5	Food consumption by children in India, 2006-2015	55
Figure 2.6	Malnutrition rates in India, 2006-2015	56
Figure 2.7	Malnutrition rates in rural and urban areas, 2015	56
Figure 3.1	Conceptual framework for smallholder farming households	60
Figure 3.2	Consumption indifference curves	65
Figure 3.3	Production possibility frontier and consumption indifference curves	67
Figure 3.4	Effect of transaction costs on consumption	69
Figure 3.5	Dietary diversity scores of women	75
Figure 3.6	Production diversity of households	76
Figure 3.7	Predictive margins of market visits	93
Figure 3.8	Nuts and seeds consumption per income quintiles	95
Figure 3.9	Dairy consumption per income quintiles	95
Figure 3.10	Vegetables consumption per income quintiles	96
Figure 4.1	Probability density functions of shocks with varying mean μ and variance σ^2	116
Figure 4.2	Effect of varying risk levels A on the optimal nutrition level	118
Figure 5.1	Distribution of patience in the world	128
Figure 5.2	Correlation of patience and income	129
Figure 5.3	Distribution of risk among gender	137
Figure 5.4	Risk levels by quintiles among households	138
Figure 5.5	Distribution of altruism among gender	138
Figure 5.6	Altruism levels by quintiles among households	139
Figure 5.7	Point elasticities of risk on income	156

Abbreviations

AIPE	Accuracy In Parameter Estimation
BMI	Body Mass Index
CHIRPS	Climate Hazard Group InfraRed Precipitation
DHS	Demographic and Health Surveys
DDW	Dietary Diversity of Women
EAR	Estimated Average Requirement
FAO	Food and Agriculture Organization of the United Nations
FGD	Focus Group Discussion
FIES	Food Insecurity Experience Scale
FOC	First Order Condition
GDP	Gross Domestic Product
GIZ	Deutsche Gesellschaft für Internationale Zusammenarbeit
GMM	Generalized Method of Moments
GPS	Global Positioning System
HDDS	Household Dietary Diversity Score
ICN2	Second International Conference on Nutrition
IRR	Incidence Rate Ratio
IV	Instrument Variable
MDD	Minimum Dietary Diversity for Infants and Young Children
MDDW	Minimum Dietary Diversity-Women
MUAC	Mid-Upper Arm Circumference
NGO	Non-Governmental Organization
NREG	National Rural Employment Guarantee
OLS	Ordinary Least Squares
PD	Production Diversity
PDS	Public Distribution System

RNI	Recommended Nutrient Intake
PPF	Production Possibility Frontier
SDG	Sustainable Development Goal
UL	Upper Limit
UNICEF	United Nations Children's Fund
VIF	Variance Inflation Factor
WDDS	Women's Dietary Diversity Score
WHO	World Health Organization
2SLS	Two-Stage Least Square

1 Introduction

1.1 Background and Motivation

The global community recognizes deficiency in dietary micronutrients as a prioritized problem. The reason is that the lack of micronutrients particularly during the first 1000 days of a life – from conception to 2 years of age – prevents individuals from developing physically and cognitively to their full potential (Kar et al., 2008). From a macroeconomic perspective the consequences are indicative, although one perspective is that undernourishment is causing poverty and has a negative effect on economic growth (von Braun, 2015b), let alone on a society's equality.

An estimated 821 million people suffer from undernourishment (FAO, IFAD, UNICEF, et al., 2018). More than 2 billion people suffer from micronutrient malnutrition, the so-called *hidden hunger* (IFPRI, 2015). The majority of food insecure people live in rural areas and 50% of food insecure people are smallholder farmers (von Braun, 2013). Around 160 million children under 5 years are stunted, meaning they are too short for their age (IFPRI, 2015). These are only a few statistics indicating that malnutrition – inadequate, unbalanced or excessive consumption of macronutrients and/or micronutrients – is a severe global problem (FAO, 2015).

Figure 1.1 on the following page shows the global distribution of micronutrient malnutrition. Considering the prevalence of anemia among women as a proxy indicator for micronutrient malnutrition, we see that Subsahara Africa as well as South Asia have the highest prevalence rates. When considering the total number of women suffering from anemia, South Asia, Southeast Asia and China as well as Brazil become the hotspots of malnutrition (see Figure A.1 on page 172 in the Appendix).

Political initiatives try to address the problem. Among others the Sustainable Development Goals (SDGs) recognize the evidence on global food and nutrition insecurity and proclaim the end of hunger by 2030 with the second SDG. Within the last decade the importance of micronutrient intake has gained more momentum among international leaders in fighting for a more food secure world. Consequently, the international community is pushing for action to alleviate micronutrient malnutrition as seen at the Second International Conference on Nutrition (ICN2) in 2014 and as ongoing large-scale initiatives demonstrate.

Figure 1.1: Prevalence of anemia among women (in percent)

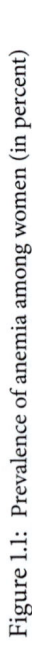

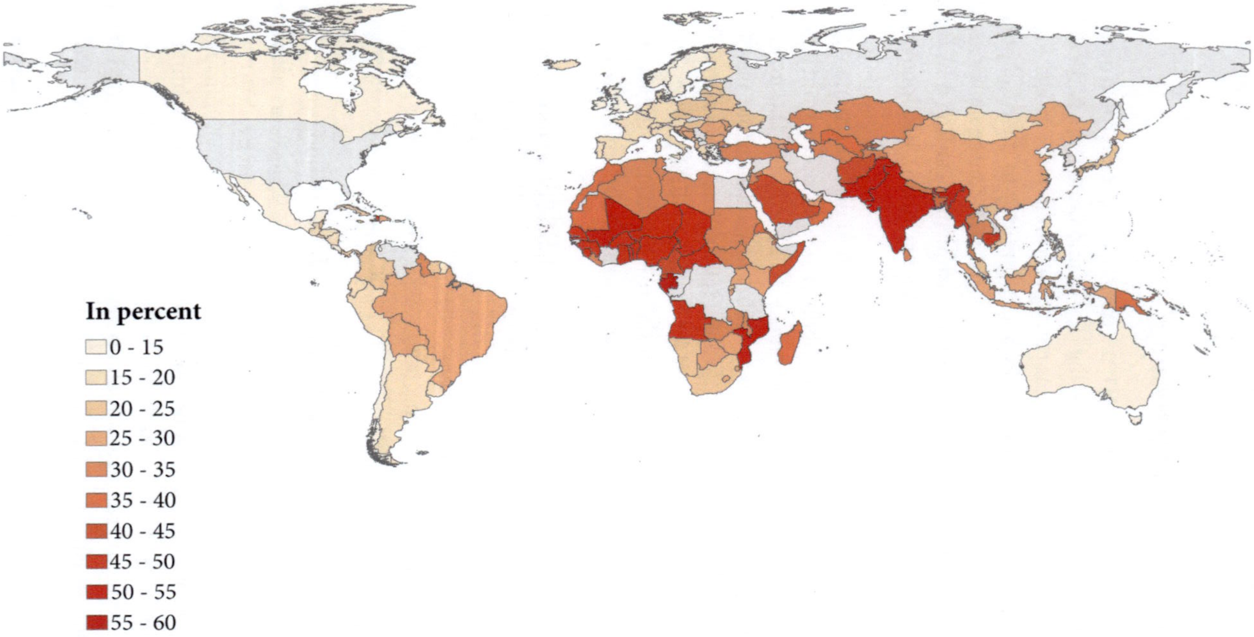

Data source: The World Bank (2017): anemic data for 2016, United Nations (2017): population data for 2015. Author's illustration

Although pushing for the same goal, each approach's theory of change is quite varied and complex. The reason is simple. Multidisciplinary research has recognized that nutrition and related health outcomes depend on a multitude of factors. With the combination of individual and cultural preferences, it is very hard to find the one solution to end hunger. However, it is possible to name the determinants that lead to a certain pattern of food consumption and to a certain state of food and nutrition security. It is possible to quantify their relevance in varying situations and it is possible to deduct policy implications for food and nutrition security particularly for vulnerable populations. This is the aim of this research.

Categories of food production

From a global perspective, food consumption is a highly heterogeneous activity, which requires basic categorization. For simplification, we differentiate between staple foods, perishable items, pulses and animal-sourced products as well as their degree of processing. First, food production data shows that staple crops such as wheat, maize and rice form the basis for most diets, although with regional differences in preference for certain cereals (FAO, 2018b). Trade statistics show the global nature of the food business. Billions of tons of cereals are shipped yearly across the globe. Some countries are production centers whereas others rely solely on food imports. Secondly, perishable items such as fruits and vegetables cannot be easily traded globally. Local production characteristics largely determine the locally available fruits and vegetables. Hence, the local production capabilities can further explain the diversification of global food consumption. Thirdly, pulses such as lentils or beans are predominantly produced and consumed in developing countries (FAO, 2016). Pulses are a protein-source that can substitute animal-sourced products with a greater ease of transportability. However, increasing income often coincides with a decline of protein consumption and with an increase of animal-sourced products, which is the fourth differentiator here. Domesticated animals or game animals have various levels of dietary importance among different populations. Dairy products or eggs are in higher demand in regions where animals are often used as a productive asset. Meat consumption is higher in regions where value is created in other economic sectors than in agriculture. A fifth factor is the degree to which processed foods are available, which varies globally. The ingredients of strongly processed foods are often difficult to identify. Convenience foods are produced for a particular taste and to minimize the time to prepare the food. On the other hand, so-called slow food contains the notion that food preparation increases the control of used ingredients accompanied with a longer preparation time. Global trends indicate that with higher income, the consumption of convenience food increases (Kearney, 2010). This trend changes the food production and processing industry, but also affects the health of consumers. All of these aspects lead to globally diverse diets. The basic categorization of food items into staple crops, vegetables, fruits, pulses and animal-sourced products helps us to understand the similarities and differences of nutritional intake.

Macroeconomic trends

Macroeconomic trends can influence individual level consumption trends. Prices of food groups over time incentivize a certain food consumption behavior. Developing countries and particularly South Asian countries such as India and Bangladesh experience a trend of diverting price changes for the past 50 years. Foods that supply dietary energy such as cereals have become relatively cheaper than foods that supply micronutrients such as vegetables and fruits (Bouis et al., 2011). It is argued that the Green Revolution and the associated research in increasing productivity of primarily staple crops contributed to a relative rise in prices of non-staple foods. This development is joined by decreasing prices of potentially unhealthy foods such as sugar or edible oils (FAO, 2018b). From an individual point of view, these prices can be considered as exogenous trends; therefore, these prices form an economic environment that incentivizes consumption of micronutrient-poor and potentially unhealthy foods.

Considering aggregated food price indices, all food prices increased globally by more than 50% in real terms in the past 15 years (FAO, 2018a). Food price indices in some developing countries rose even more than non-food consumer price indices (Bouis et al., 2011). Low income groups in e.g. India and Bangladesh have a comparatively high share of food expenditures to their available budget. Increasing food prices amplify this situation further (Deaton, 2003). The general trend of increasing food prices additionally limits the consumption of non-food goods by low income groups. To mitigate the limitation and since non-staple foods are often regarded as dispensable, consumption of these foods is often reduced.

Contribution

Following the considerations above, this research focuses on diets of vulnerable groups in rural environments that face an increased likelihood of shocks. Vulnerable groups are here understood as households with low income and particularly women of reproductive age, as well as children below 5 years of age. This research studies the factors that influence food consumption patterns: production choices, markets and preferences.

The next section of this introductory chapter presents the research objectives of the specific analytical chapters. A conceptual framework is introduced in Section 1.3 that explains the linkages among essential aspects of food and nutrition security and that guides the structure of this research. In Section 1.4 we present the current state of information on nutrition and provide the foundation for the research rationale, why and how food consumption proves to be vital for individual and societal development. Although nutrition is a very well-studied area, it still lacks precision in its measurement. Section 1.5 therefore discusses our approach to focusing primarily on nutrition intake and how we intend to measure it. The research design for the primary data, which is the core dataset for the empirical analysis, is presented in Section 1.6 and complements this introductory chapter.

1.2 Research Objective and Questions

The research objective overall is to identify factors that influence consumption choices and to estimate their effects on food and nutrition security. These factors are related to production, markets and individual preferences. Various pathways that lead to certain food consumption choices are teased apart. One descriptive chapter and three analytical chapters work towards the research objective by addressing specific research questions.

The descriptive Chapter 2 - *Focus on India: Food Prices and Food Security* - presents the macroeconomic environment in India and discusses in more detail the effects of price trends on the food and nutrition security situation. The latest data for macroeconomic indicators, food consumption and prices are presented and interpreted in line with current malnutrition rates.

Chapter 3 - *Production Diversity and Diets* - considers production choices of smallholder farmers that influence nutrition intake. Smallholder farming households that at least partially engage in on-farm activities tend to consume part of their produce within the household. Accordingly, the choice of diverse production also potentially affects the choice of diverse consumption. It is, however, a pathway that might decrease in relevance the more market-integrated a household is. Hence, the more production choices are influenced by markets, the more consumption choices are enabled through markets. The research literature has discussed these linkages and provided empirical evidence. The contribution of this chapter to the existing literature is threefold: (1) proposing a theoretical link for the effects of production diversity on dietary diversity, (2) introducing a single instrument variable as identification strategy, and (3) applying a coherent methodology for estimating production and dietary diversity in the Indian context. The research questions are:

1. Do production choices affect nutrition choices of smallholder farmers and if so, to what extent?
   Hypothesis:
   Increasing on-farm production diversity along the production possibility frontier increases the dietary diversity of the household and of individual household members.

2. Does market access influence this relationship?
   Hypothesis:
   The consumption of foods and dietary diversity increases with market access, and, at the same time, the relationship between on farm production diversity and dietary diversity diminishes.

Chapter 4 - *Considering Preferences for Food Consumption* - focuses on economic preferences and dietary intake. Many underlying and basic determinants frame consumption choices. Also, individual preferences form to some extent the decisions on what to eat. The relation between preferences and economic behavior has been widely studied, although the connection to individual diets and food and nutrition security is lacking. We contribute to the literature by demonstrating that deep preferences such as risk preference, time preference and altruism directly affect individual nutrition. In this chapter, we develop a theoretical model to lay the conceptual foundation for these linkages. We are embedding the model in the current literature and consider the situation of populations that are at risk of undernutrition. The research objective and contribution is to present an economic model that predicts the effects of preferences on current nutrition of individuals.

Chapter 5 - *Effects of Preferences on Food Consumption* - continues the consideration that derive from the previous theoretical elaboration with a focus on risk preference and altruism. We are applying the model in the Indian setting. A survey tool is developed that enables the data collection of preferences on the individual level in combination with socioeconomic and nutrition information on the household and village level. Hypotheses that are derived from the theoretical model are tested and the effects of preferences on nutrition intake estimated. The results indicate highly significant and robust linkages between preferences and nutrition. We answer the following research questions:

1. Does risk preference affect dietary intake and if so, to what extent?
   Hypothesis:
   Risk preference measured as an individual level of risk aversion positively affects the current nutrition level of individuals.

2. Does altruism affect dietary intake and if so, to what extent?
   Hypothesis:
   Altruism of another household member j towards an individual i positively affects total utility of i's nutrition due to the increased amount of food items available to i.

1.3 Conceptual Framework

The field of food and nutrition security is increasingly cross-disciplinary, focusing on nutrition outcomes and on effects on the environment. Existing conceptual frameworks are similarly quite diverse, although two core frameworks prevail: firstly, the Food and Agriculture Organization of the United Nations (FAO) food security framework, which is a basic yet comprehensive set of concepts and for this reason widely used (FAO, IFAD, and WFP, 2015); secondly, the conceptual framework of the determinants of nutritional status from United Nations Children's Fund (UNICEF), which is a framework that guides project implementation and research alike (UNICEF, 1990). The classification of this research will not be possible without introducing these two frameworks, which is the aim of this section. In this section we also present the overall conceptual framework that guides this research. Each analytical chapter further discusses the literature and additional theoretical concepts that are more precise and suitable to each specific research question.

FAO Food security framework

The FAO framework depicts four dimensions of food security: food availability, economic and physical access to food, food utilization and stability over time (FAO, IFAD, and WFP, 2015). Thus, food security exists when „all people, at all times, have physical, social and economic access to sufficient, safe and nutritious food that meets their dietary needs and food preferences for an active and healthy life" (FAO, 2015, p. 53). In this framework, *food availability* considers the sufficient amount of foods that are either stored or produced, or that can be transported, for instance, through trade. *Food access* considers the economic purchasing power for an adequate quantity of food. In extreme cases, food access can also be provided through food aid, hence, without any economic access but through physical access provided externally to an individual. *Food utilization* considers handling of food either post-harvest, during storage or at time of preparation for consumption. More frequently, food utilization emphasizes so-called "nutrition-sensitive" practices that can optimally improve the nutritional-benefit of food consumption or limit the nutritional loss. These include, for instance, feeding practices of children or water and sanitation hygiene practices. *Stability over time* emphasizes the aspect that a sufficient nutrition is needed at all times, hence, that also short periods of malnutrition have detrimental effects.

The FAO food security framework presents basic principles that have undergone many extensions since its initial formulation at the 1974 World Food Conference. At the intergovernmental level, guidelines are deducted from the framework that guide many interventions on food and nutrition security. The most recent internationally adopted Global Strategic Framework for Food Security and Nutrition presents a synthesis of existing concepts and best practices to provide recommendations for coherent actions in that it "provide[s] an overarching framework and a single reference document with practical guidance on core recommendations for food security and nutrition strategies, policies and actions" (CFS, 2017). Linkages that are focused on nutrition - such as the Global Strategic Framework presents on the basis of the FAO food security framework - are conceptualized more comprehensibly at the UNICEF framework for nutrition.

UNICEF Framework for nutrition

The UNICEF framework (see Figure 1.2 on the following page) displays linkages and interlinkages between various determinants, respective causes of malnutrition and the resulting nutrition outcomes (UNICEF, 1990). Various elements of the FAO food security framework can be recognized as underlying determinants (e.g. food availability and access under food security resources). Also, the goal of food security is similar to an active and healthy life, depicted as nutritional status that reflects weight, height or, generally, the possibility to fully develop and sustain one's physical and mental abilities. The UNICEF framework adds value to the nutritional debate by describing the linkages between basic, underlying and immediate determinants with the eventual nutritional status.

Basic determinants reflect the societal structures and processes that present environmental conditions for individuals or households, which are usually exogenous to them. *Underlying determinants* are more household-specific in regard to individual capabilities, economic viability, or utilities. *Immediate determinants*

Conceptual Framework

Figure 1.2: UNICEF conceptual framework of the determinants of nutritional status

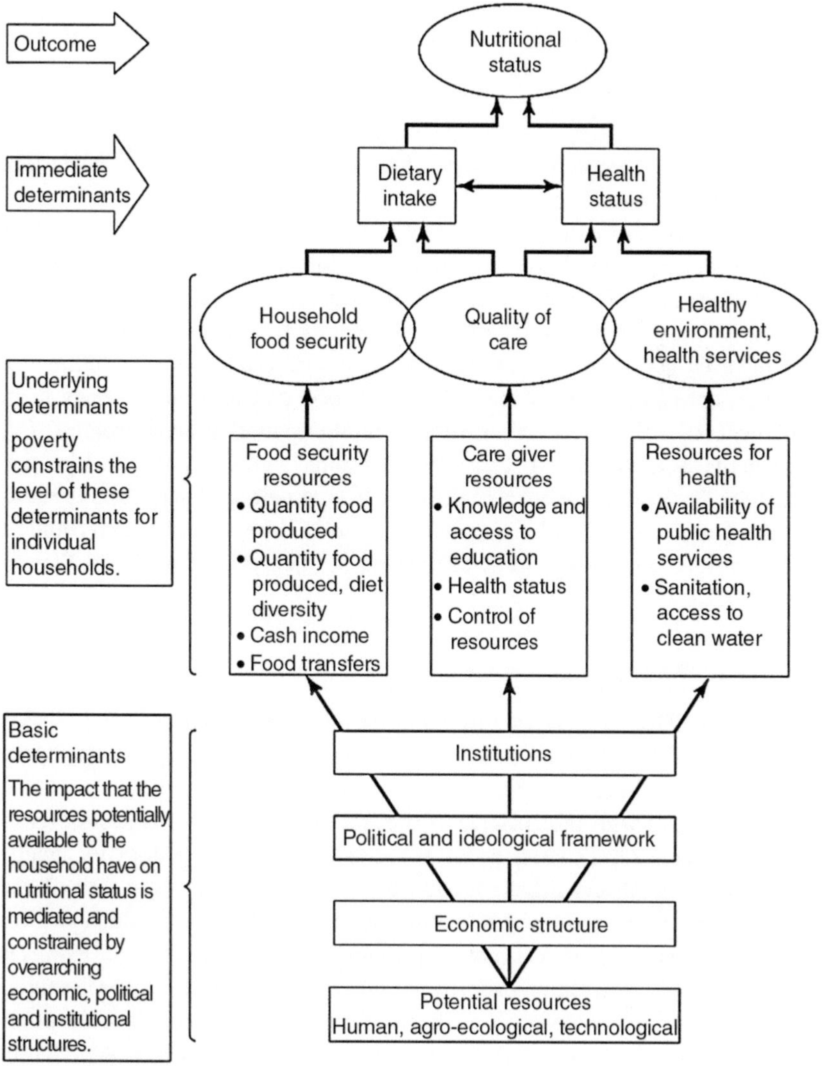

Outcome → Nutritional status

Immediate determinants → Dietary intake ↔ Health status

Underlying determinants

poverty constrains the level of these determinants for individual households.

- Household food security
- Quality of care
- Healthy environment, health services

Food security resources
- Quantity food produced
- Quantity food produced, diet diversity
- Cash income
- Food transfers

Care giver resources
- Knowledge and access to education
- Health status
- Control of resources

Resources for health
- Availability of public health services
- Sanitation, access to clean water

Basic determinants

The impact that the resources potentially available to the household have on nutritional status is mediated and constrained by overarching economic, political and institutional structures.

- Institutions
- Political and ideological framework
- Economic structure
- Potential resources: Human, agro-ecological, technological

Source: Pasricha and Biggs, 2010

are direct results in the form of actual dietary intake and health status. The outcome of all determinants is represented as nutritional status.

The UNICEF framework is of importance in that it differentiates the various levels of determinants. Food and nutrition security is a multi-factorial task. Changes in one sector that possibly aim at improved nutritional status are likely to affect other sectors in a detrimental or synergistic way. Hence, this framework calls for a more comprehensive view of undernutrition than merely the lack of food or the access to it.

In an addition to the framework, UNICEF and others recognize that immediate determinants have short-term and long-term consequences on the nutritional status (UNICEF, 2015). Short-term consequences can be increases in risk of mortality and morbidity, whereas long-term consequences are seen in impaired cognitive abilities, reduced economic productivity or future risk for non-communicable diseases. Moreover, food and nutrition security is an intergenerational matter; malnutrition of mothers (more general, women of reproductive age) can directly affect the nutritional status of young and unborn children. Feedback loops are recognized as well. A deteriorating nutrition status might affect basic and underlying determinants in the long-run, which could result in poverty-increasing vicious circles (Black, Allen, et al., 2008; Black, Victora, et al., 2013).

Conceptual framework

The presented frameworks are commonly used for various purposes and reasons. Yet, each food and nutrition security framework needs to comply to a set of principles without which its relevance might be undermined. Von Braun discussed these principles with the example of the FAO food security framework (compare von Braun, 2015a). First, the FAO food security framework does not include causal linkages between drivers and impacts on nutrition outcomes. Second, the determinants are conceptualized as independent of each other, although positive or negative synergies are existent. Third, dynamics are not included. However, time dependent agricultural patterns or vulnerabilities to shocks are relevant to sustained nutrition outcomes. Fourth, broader contexts particularly in the political dimension need to be specified. Institutions and governance can have relevant impacts on nutrition outcomes, which are only marginally addressed in the predominant frameworks. Accordingly, a suitable conceptual framework for this research should consider justified criticism to established concepts and at the same time be operational (for example, Jaenicke and Virchow attempted to create one for food and nutrition security policies, see Jaenicke and Virchow, 2013).

Figure 1.3 on the following page visualizes the guiding conceptual framework that is designed for this research. The conceptual framework reflects in general a decision making process that is dynamic and highly interlinked. Available, accessible and correctly utilized nutrition depends primarily on the choices that each individual makes given an accessible food basket in the market. Market is here understood as a broad concept that entails not only physically available products, but also pricing mechanisms, which influence individuals' purchasing behavior as well as possible production choices. Two broad categories that are interlinked with each other are supply and demand dimensions. Similarly, we deliberately do not disassociate supply and demand dimensions, but consider supply and demand as rather broad concepts that form individual decisions but are also influenced by individual decisions. Each aspect of these dimensions are interlinked and need to be considered in interaction rather than separately. Generally, the whole decision making process is embedded in broader environments, which are of economic, political and ecological nature. These environments frame the basic conditions for the dimensions but are also primary stimuli over time, in the sense of shocks or regular fluctuations such as agricultural seasons. The framework is read as a sequence from left to right reflecting a time dimension with an eventual feedback loop from the nutritional status to future consumption choices.

The conceptual framework for this research utilizes the two presented frameworks of FAO and UNICEF by adjusting these for our purpose and by integrating the criticism. It overcomes some of the other frameworks' limitations and proves suitable to this research. This research turns the focus towards households in rural environments and their diets. Decisions about diets are made consciously and unconsciously on a household as well as individual level. Ultimately, however, the decision is formed on the basis of which food items are available and accessible and - most of the time - on what the household wants to consume. Accordingly, supply and demand reflect household decisions in the proposed framework. The UNICEF framework recognizes the multitude of factors leading to a certain nutrition status. This is reflected in the proposed framework as well. Additionally, looking at rural environments that depend on agriculture as a means of livelihood, food supply and food demand are interlinked with a varying strength depending on the level of market integration of a food producer. This can be seen in line with the agricultural household model's non-separability property for the choice of food items: consumption decisions depend on the production decisions when market linkages are limited (Taylor and Adelman, 2003). The ecological, economic and political environments present not only a setting that is exogenous to the household; shocks and trends are also often induced and mitigated

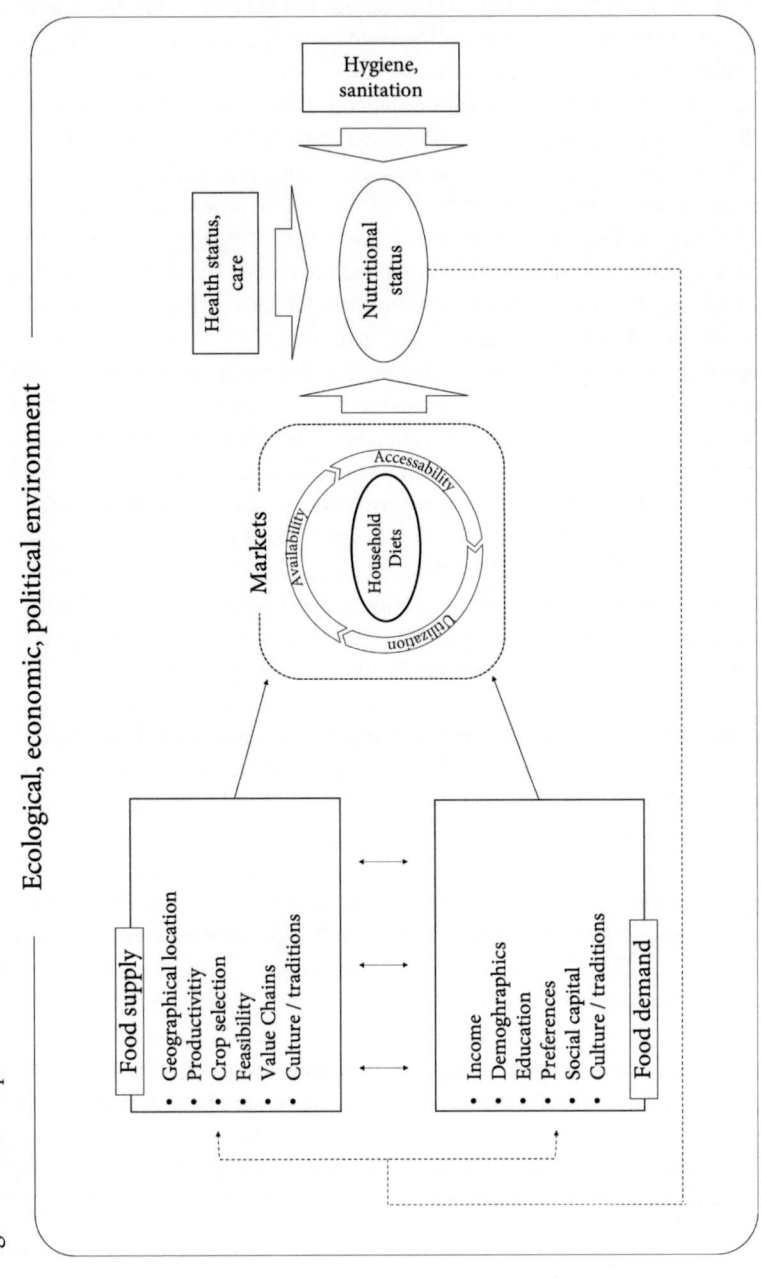

Figure 1.3: Conceptual framework

by these environments. Therefore, shocks are specific and can occur for each aspect of the entailed concepts. Covariate shocks and seasonality will have their roots, for instance, in macroeconomic crises or in climatic seasons. Idiosyncratic shocks might affect only the income ability of a household or the health status of an individual. For the purpose of this research, it is crucial to understand a risky environment as an environment that can positively, but mostly negatively affect each dimension of the comprehensive decision making process for diets.

1.4 Nutrition – Engine of Development

The concept of nutrition needs further explanation. By following the literature in nutrition science, this section explains the basics of nutrition and corresponding definitions for food insecurity. We will also hint to the importance of nutrition for development.

Essentially, everyone is a machine that needs energy to function. Food is the fuel (in combination with other forms of energy such as sun light). Nutrition is the combination of food items that are consumed. Diets are characterized in terms of quantity and quality, in terms of macronutrients and micronutrients. Macronutrients – that are also called energy-supplying nutrients – contain proteins, fats, and carbohydrates. The majority of micronutrients are not energy-supplying, but guarantee essential roles for the body to function, such as metabolism or the immune system. Micronutrients are primarily vitamins and minerals. Macronutrients and micronutrients vary in the amount in which they are required. The requirements themselves are again dependent on age, physical appearance, gender, activity level and other factors. For demonstration purposes, daily consumption of macronutrients tends to be more than 500g, the micronutrients Vitamin C around 100mg and Vitamin B12 only 5µg. Macro- and micronutrients have in common the fact that the lack of both can lead to death, although for most micronutrients it holds true that only a sustained undersupply of these will lead to death. Health impairments or reduced physical and cognitive impairments are more often the consequence of micronutrient deficiencies (Gibney et al., 2002).

Two additional facts lead us to an understanding of why increasing attention is put on micronutrients. A lack of macronutrients can be recognized by an individual as a sensation of hunger. A sufficient supply of energy will reduce the hunger accordingly. A lack of micronutrients, however, does not cause an equivalent sensation. Instead, after some time of a deficient supply of micronutrients, symptoms such as muscle pain or exhaustion occur, which cannot directly be

Figure 1.4: Criteria for risk of deficiency and excess food intake

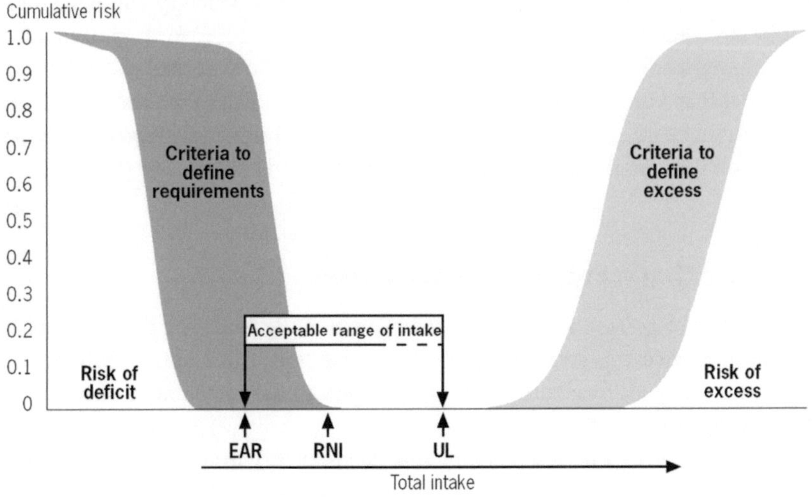

Source: FAO and WHO, 2004, p. 3

linked to the lack of a certain micronutrient by an individual. Therefore, micronutrient deficiency is also called *hidden hunger*. Secondly, certain micronutrients only exist in a few food groups such as organ meat or fish, but are still essential: e.g. Vitamin A, D, B12, iron, iodine, and zinc. Other micronutrients have to be consumed constantly because the body cannot store these: e.g. Vitamin C. Thus, a sustained balanced diet or, in other words, a good quality of diet, is difficult to achieve but nevertheless imperative for the full development and health of an individual (Biesalski, 2013).

Most countries publish national guidelines detailing the specific locally-available food items with which the requirements for a balanced diet can be met daily. However, access and availability to these food items, let alone the preference for them, differ between individuals and households. Figure 1.4 above defines the intake levels that are usually considered for under- versus overnutrition. The Estimated Average Requirement (EAR) is the average daily intake required for 50% of the population. The Recommended Nutrient Intake (RNI) is defined as EAR plus 2 standard deviations, thus meeting the needs of 97.5% of the population. The Upper Limit (UL) depicts the point up to which no evidence of toxicity can be found. Essentially, the recommendation by World Health Organization (WHO) and FAO is to consume within the range of the RNI and UL in order to have a healthy diet (FAO and WHO, 2004, p. 5).

Figure 1.5: The cycle of micronutrient inadequacies

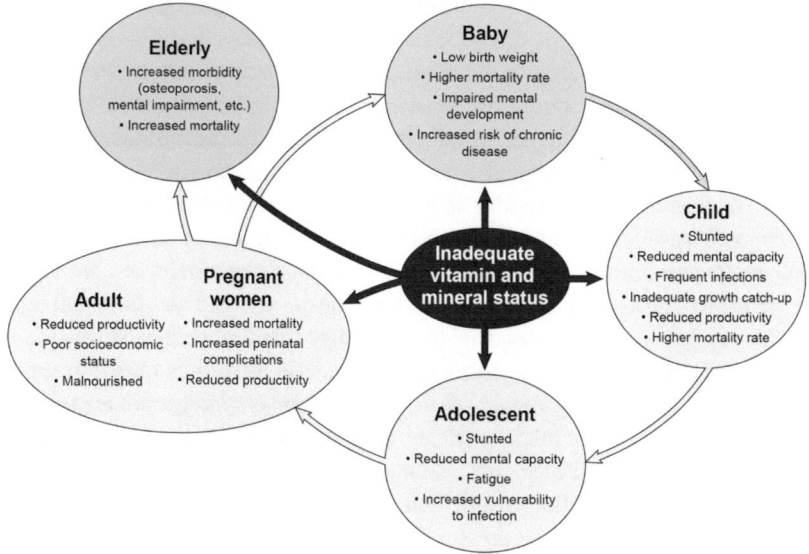

Source: Bailey et al., 2015, p. 26

The focus of nutrition research is often on women and children. The literature shows that during pregnancy and the first years of life, a lack of nutrients has a severe effect on children. This holds true for both macronutrients and micronutrients. A balanced nutrition in these early years is necessary for brain development, physical development and the immune system. Even lifelong health can be affected because of early undernutrition. Through the pregnancy a malnourished mother can give birth to a malnourished child; malnutrition is inheritable. Consequences are manifold depending on the lacking micronutrient, but each consequence amounts to preventing an individual from developing to her full capabilities (Biesalski, 2013). Figure 1.5 above depicts the cycle of malnutrition. In the Appendix on page 173 you can find a list of consequences for mother and child due to micronutrient deficiencies. Accordingly, this research focuses primarily on women and children below 2 years of age (1000-days window), respectively, on households that have these members.

Nutrition not only affects individuals, but also has a much larger impact on the development of societies and – in an even larger sense – species. Macroeconomic consequences of undernutrition are seen in a lower productivity of labor due to lower skill levels and lower physical abilities, higher rates of non-communicable diseases, and lower fertility rates. Furthermore, Biesalski remarks that "the diversity of food as a foundation of a balanced supply of micronutrients has been a driver of human development and is most probably still today" (Biesalski, 2015, p. 236). The understanding of nutrition needs to be broadened; it is not only a necessity for survival but much more an engine of evolution. Cognitive and physical abilities are dependent on the supply of nutrients, but evolutionary changes can also be induced by a limited or excessive supply of nutrients.[1]

In conclusion, a few essential definitions can be introduced. One speaks of *malnutrition* as inadequate, unbalanced or excessive consumption of macronutrients and/or micronutrients on the basis of the dietary requirements. *Stunting* – too short for the age, *underweight* – too light for the age, and *wasting* – too light for the height – can be consequences of prolonged *undernutrition*, starvation is its most extreme form. These consequences are also termed as nutrition status or nutrition outcome. *Micronutrient deficiency* or *hidden hunger* are terms used for the particular lack of micronutrients. *Overnutrition* is the state of consuming more than the required nutrients, leading to overweight and obesity. Equipped with a better understanding of nutrition and with the core concepts defined, we want to turn the focus on the measurement of nutrition.

1.5 Measurement of Nutrition

As Barrett recognizes on the measurement of food and nutrition security: "measurement drives diagnostics and response" and research is aiming "to improve the disaggregated identification of food-insecure sub-populations and their targetable characteristics and behaviors" (Barrett, 2010, p. 827). Accordingly, the measurement of nutrition needs on the one hand to operationalize the concepts of food and nutrition security and to take into consideration the status quo of nutrition science as stated above. On the other hand, the measurement might aim to satisfy at least the characteristics of good indicators in order to be usable on a

---

1 One notable example for excessive supply is the human brain development that was probably triggered by a higher energy supply and micronutrient supply through new food items found in new habitats. An example of the effect of limited supply is the depigmentation of skin, which possibly resulted due to an improved Vitamin D synthesis. Skin pigmentation limits UV light penetration, which is necessary for Vitamin D synthesis in the human body. Therefore in areas of relatively less UV light intensity (such as in European habitats) lighter skins might have had an evolutionary advantage. Furthermore, today Vitamin D deficiency can be found in almost all parts of the world due to full body clothes (Biesalski, 2015).

larger scale: specific, measurable, attainable, realistic and time bound (compare Doran, 1981).

The categorization and identification of commonly used food and nutrition security indicators can be found in various publications (Cafiero et al., 2014; Herforth and Ballard, 2016; Pangaribowo et al., 2013). The indicators that are most often used in research are categorized in measuring (1) biochemical status, (2) anthropometry, (3) diet and food consumption, and (4) food access (Herforth and Ballard, 2016).

Biochemical indicators

The micronutrient content in human tissues presents the most exact measurement and, given an individual requirement for each micronutrient (compare Figure 1.4 in the previous section), the extent of malnutrition can be stated (Bailey et al., 2015). The measurement methods are quite elaborate; technical implementation, ethical concerns and individual consent are all reasons why biochemical indicators are not frequently used and are limited to resource-rich projects. Furthermore, biochemical assessment is mostly limited to iron, Vitamin A, zinc, iodine and B12. The methods for these micronutrients are currently most cost-effective and particularly iron and Vitamin A can guide as proxy indicators for other micronutrient deficiencies in children and adults alike. But even though biochemical indicators present the gold standard from a research perspective, we cannot use these in this research due to budget and time constraints.

Anthropometric indicators

Anthropometric indicators present the nutrition outcome by measuring weight, height and age of individuals. In regard to children, population-specific reference values are used to calculate z-scores for underweight (weight-for-age), stunting (height-for-age), and wasting (weight-for-height). One generally speaks of these forms of undernutrition whenever the z-scores are below -2 standard deviations. Regarding adults and specifically women of reproductive age, often the Body Mass Index (BMI) is measured to determine the nutrition status. The BMI represents the weight per m^2. The reference defines underweight as a BMI below 18.5, normal weight between 18.5 and 25, and overweight above 25. For children and women alike, the Mid-Upper Arm Circumference (MUAC) is another often used indicator. The measurement of the MUAC is a quick but imprecise measurement for the nutrition status. It is mostly used in cases where large numbers of individuals need rapid medical attention, such as during famines.

This research considers anthropometrics as an additional variable for robustness checks. The hypotheses do not aim at the nutrition outcome level, but nutrition intake. Referring to the UNICEF framework of nutrition (compare Figure 1.2 on page 31), nutrition outcome as measured by anthropometrics represents another level that is influenced by a multitude of factors that require in large part constant conscious decisions by an individual[2]. In the tradition of behavioral economics, we see the individual's decision making process within bounded rational choices. We also understand that food consumption is an essential activity, which provides a direct feedback to the individual so that consumption choices on food intake are more likely to be constantly conscious than e.g. nutrition-sensitive activities. Therefore the hypotheses will aim at food consumption and not directly at anthropometrics.

Diet and food consumption indicators

Diet and food consumption indicators measure the variety and quantity of food intake of individuals or households. Different food items are categorized in certain food groups, the consumption of which forms the basis for calculating each indicator. The most common dietary intake indicators are the Women's Dietary Diversity Score (WDDS), the binary form Minimum Dietary Diversity-Women (MDDW) as well as its count form Dietary Diversity of Women (DDW), the Minimum Dietary Diversity for Infants and Young Children (MDD) for young children, the Food Consumption Score for households and various micronutrient-focused indicators such as Vitamin-A rich food intake. The indicators differ primarily in the composition of food groups, e.g. the MDDW uses 10 food groups, whereas the WDDS uses only 9 food groups and the MDD uses 7 food groups. The various food groups per indicator are listed in the Appendix on page 174. The indicators are surveyed by requesting the respondent (or her mother) to recall all the consumed foods for the past 24 hours. Only if a certain amount of a given food item is consumed in the past 24 hours will it be counted as consumed.

Food consumption indicators are similar in that these tend to measure the micronutrient (mal)nutrition. These are in this sense proxy indicators, which are easier to survey than biochemical indicators. The most current and generally accepted food intake indicator - the MDDW that postulates a micronutrient sufficiency on population-level if at least 5 out of 10 food groups are consumed -

---

2 On the correlation between nutrition intake and anthropometrics compare Arimond and Ruel, 2004; Jones, Ickes, et al., 2014; Martin-Prével et al., 2015.

displays the imprecise character (compare FAO and FHI 360, 2016): A five-year research project that "carefully and extensively reviewed and also discussed against non-technical criteria for evaluating indicators" eventually proposed the MDDW. It is a proxy indicator for micronutrient nutrition that reaches a mean probability of micronutrient adequacy of 60% in regard to biochemical reference points (Martin-Prével et al., 2015, p. xiv). This limitation holds also for the other indicators; hence, interpretations in regard to actual micronutrient malnutrition need to be cautious when discussing food consumption indicators.

Food access indicators

Food security measures in the access dimension are nutrition indicators that indicate access to food or to adequate nutrition. The methodologies for these indicators are quite diverse, yet around 80% of research projects in the field of nutrition utilize these (Herforth and Ballard, 2016). Among these, two are often used: the Household Dietary Diversity Score (HDDS) and the Food Insecurity Experience Scale (FIES).

The HDDS is technically similar to a food consumption indicator in that it inquires the food consumed by a household in the past 7 days considering 12 food groups (the food groups are listed in the Appendix on page 174). Given the aggregation on the household-level and considering the sum of 7 days, this indicator does not reflect any micronutrient adequacy but is much more of a proxy indicator for access to food (Swindale and Bilinsky, 2006). The FIES is a subjective measure of food insecurity (Ballard et al., 2013). A household or individual is asked to respond to 8 questions on food security that cover a recall period of up 12 months, e.g. "Did you eat less than you thought you should because of a lack of money or other resources?" (all questions are listed in the Appendix on page 176). Whereas the HDDS reflects a high number as a positive measure for food security, the FIES considers a low number as food secure. Since both indicators measure the same dimension for food security, a high negative correlation (close to -1) is expected and can guide as a robustness check.

In selecting the main variables that quantify the concept of nutrition, we follow Verger's argument: "In the context of studies of linkages between agriculture, markets and food consumption, whether the objectives are to improve human nutrition, sustain productive ecosystems, ensure economic development or a combination of these, great care should be exercised when selecting and interpreting metrics. In order to allow comparisons across studies, regions or countries, it is crucial that standardized dietary diversity scores, accepted by the international scientific community, are used" (Verger et al., 2017). In 2013, Turner et al.

mapped current and future research on nutrition-sensitive agriculture among others to identify research gaps (Turner et al., 2013). At this time, the most commonly used indicators were food consumption indicators used by 93% of the evaluated research projects, food access indicators used by 80%, and anthropometrics were used by 72% (compare also Herforth and Ballard, 2016, p. 3). This research follows the established methodology for estimating individual, household and population malnutrition in that we primarily use the DDW, HDDS, consumption of particular micronutrient-rich food groups and anthropometrics for robustness checks. Furthermore, we also utilize two indicators that were published after the overview article by Turner et al. (2013), but have gained wide acceptance in the literature, which are the MDDW and the FIES (Leroy et al., 2015). Hence, the focus is on food intake emphasizing micronutrients and differentiating between household-level and individuals, i.e. women and children. These food intake indicators present the main dependent variables for the analysis in this research.

1.6 Research Design

1.6.1 Study sites

The last section of this introductory chapter presents the research design for the surveys, which are used for the analysis. The research draws its empirical findings primarily from an observational study that was conducted from January to July 2017 in three regions of India. The study sites were selected based on four criteria: (1) malnutrition had to be prevalent, (2) rural areas where agriculture is the primary source of income for the majority of households were chosen, (3) they are furthermore rural areas where environmental shocks are occurring, and (4) they were accessible within the financial and time constraints of this research.

The first region was selected in the state of Jharkhand in collaboration with Welthungerhilfe India. Jharkhand is one of the poorest states with devastating figures for malnutrition. Only 24% of the population lives in urban areas, and, given its geographical location, Jharkhand is prone to droughts. It was possible to conduct the survey in a remote area of Jharkhand with the support of the local Non-Governmental Organization (NGO) Pravah. Pravah supported the selection of enumerators and prepared access to the selected villages for the survey.

The second region was selected in West Bengal of east India with support by the University of Calcutta. West Bengal represents more an average state of India in terms of socioeconomic and nutrition statistics. A study site was chosen that borders on Bangladesh, being separated from it by the mighty river Padma. This area is quite fertile, thus agriculture provides a large portion of the population's opportunities for income. However, the closeness to the river also presents the risk of floods, which are occasionally life threatening to the population. With organizational support by the University of Calcutta and with experienced data enumerators, it was possible to collect data in this area.

The third region presented itself as an opportunity through collaboration with the Deutsche Gesellschaft für Internationale Zusammenarbeit (GIZ) in Karnataka in Southern India. GIZ implements a project on agricultural innovations and was in need of a baseline study for their project. GIZ offered to finance the data collection in Karnataka while respecting the conditions needed for this research, in exchange for usage rights to the collected data.

Karnataka is a region that is economically slightly above average in India. Food security in terms of food access is not a prioritized issue, however, micronutrient malnutrition rates are high. The chosen study site is approximately 1000m above sea level and comprises a different agroecological zone than Jharkhand and West Bengal. In this area, the focus is on cash crops, with agriculture providing the major source of income. GIZ helped in selecting data enumerators and provided the logistics for the data collection.

Jharkhand and West Bengal are logistically easy to reach within one day of travel from Kolkata, the location in which the affiliated University of Calcutta and the researcher were residing. Local support at the study sites enabled the data collection over the period of three months in each region. Support was provided by district officials, by village leaders for each selected village and by Anganwadi workers[3].

To provide a better understanding of the food and nutrition security situation as well as other socioeconomic indicators, Table 1.1 on the following page presents some characteristics of the regions.

---

[3] Anganwadi are rural child care centers financed by the Indian government under the Integrated Child Development Service Program. The Anganwadi workers among others provide nutrition and health information, collect nutrition data of children and support child care practices.

Table 1.1: Characteristics of rural study regions

	Jharkhand	West Bengal	Karnataka	India
General characteristics				
Rural population in millions	25.0	62.1	37.5	833.7
Rural population (%)	76.0	68.1	61.3	68.9
Employment in agriculture (%)	65.6	62.2	51.2	49.6
Below national poverty line (%)	40.8	22.5	24.5	25.7
Women who are literate (%)	51.5	66.9	63.8	61.5
Men who are literate (%)	75.9	79.7	81.2	82.6
HHs with electricity (%)	74.4	92.0	97.0	83.2
HHs with sanitation facility (%)	12.4	45.5	42.6	36.7
Under-five mortality rate[a]	58	38	38	56
Food and Nutrition Security				
Stunted children[b] (%)	48.0	34.0	38.5	41.2
Wasted children[b] (%)	29.5	21.6	26.9	21.5
Underweight children[b] (%)	49.8	33.6	37.7	38.3
Underweight women[cd] (%)	35.4	24.6	24.3	26.7
Underweight men[d] (%)	25.6	20.3	18.4	23.0
Overweight women[cd] (%)	5.9	15.0	16.6	15.0
Overweight men[d] (%)	7.5	11.2	17.1	14.3
Anemic children[b] (%)	71.5	53.7	63.3	59.4
Anemic women[c] (%)	67.3	64.3	46.2	54.3

[a] Per 1000 live births

[b] Children age 6-59 months

[c] Women age 15-49 years

[d] Measured as BMI below normal (BMI < 18.5 kg/m^2) respectively above normal (BMI $\geq$ 25.0 kg/m^2)

Data source: Government of India, 2013, 2018b; IIPS and ICF, 2017

1.6.2 Sampling

The data for this research comes from a cross-sectional household survey that is amended with village surveys and market surveys as well as with the recall of time-variant information for the past 5 years (on household level) and for the past 20 years (on village level). A similar survey design is used in each study site.

Sample size

Sample size considerations are based on the goal to narrow the confidence interval for estimated dietary diversity scores, which represent the main dependent variables in the empirical analysis. Accordingly, the Accuracy In Parameter Estimation (AIPE) is the underlying methodology for the sample size calculation (Maxwell et al., 2008). A simple random sample size can then be calculated as (according to Levy and Lemeshow, 2008, p. 74):

$$n \geq \frac{z^2 N P_y (1 - P_y)}{(N-1)\varepsilon^2 P_y^2 + z^2 P_y (1 - P_y)}$$

with n being the sample size, z is the reliability coefficient, N is the population size, P_y is the true unknown population value and ε the maximum relative differences allowed between the true unknown population value and the estimate. Considering dietary diversity as a score from 0 to 10 food groups, local key informants suggested that P_y is approximately 4 in marginalized rural areas of East India. We further set ε as 0.05, which represents a difference of 0.1 food groups that we allow between the estimate and the true value considering a two-sided confidence interval; the population size is set to infinite and we use 1.96 for z to represent 95% confidence. The calculation solves to a rounded sample size of 384. Keeping in mind the limit for overfitting the regression model later on, we can use approximately less than 40 independent continuous variables in a multiple regression equation considering dietary diversity as dependent variable (Rothman, 2012, p. 226). The actual sample sizes per region are chosen to exceed the calculated sample size in order to take into account possible issues during the data collection.

Sampling design

Different sampling designs were applied, depending on the local opportunities and feasibilities. Sampling frames were predefined per region. Stratified random sampling was used in order to have a sampling in each region that is representative on the village level while also being representative for the sampling frame. Still, the sampling design differed in the various regions.

In Jharkhand, 49 villages in proximity to each other received a census by the local NGO Pravah in 2016. These 49 villages are all neighboring to each other, and the only selection category of these 49 villages to be included in the census was the location in a predefined area. The sampling frame for Jharkhand accordingly consisted of these 49 villages. Stratification took place on the level of household members. One group of interest are children under 2 years of age. As this group tends to be the most difficult to target given the relatively small size, the sampling criteria is set as households with at least one child below 2 years of age at the time of the survey. Between the local census and the time of the survey, up to 6 months had passed. In order to possibly also include in the sample those households in which a child has been born during this time, we aimed at including *all* households in the survey that had a child below 2 years. The effect on the sample size was acceptable and feasible; thus 490 households in 49 villages were included.

In West Bengal, Murshidabad district was selected as a survey area with a population of approximately 5.86 million (Government of India, 2018b). The survey was limited to the Block Bhagwangola-II within the subdivision Lalbagh with a population of approximately 158,024 households as of the 2011 Census. Within Bhagwangola-II, four Gram Panchayats were selected on the basis of their location. These Gram Panchayats combine approximately 23,763 households in 35 villages (as of the 2011 Census). The sampling frame accordingly consisted of all households with at least one child below 2 years of age in these 35 villages. The sample size was set to 402, creating a proportionate number of households per village as per the 2011 Census total number of households. Stratified random sampling was achieved by applying the random walk technique in the villages that were previously selected (United Nations, 2008). Figure A.2 on page 179 in the Appendix shows a map of the surveyed households in Jharkhand and West Bengal.

In the region of Karnataka, the sampling design was different from Jharkhand and West Bengal due to the cooperation with GIZ that was intended to create a baseline for future impact assessment in 2019. The region was predefined by GIZ, which is a region in which project activities on agricultural trainings for potato production would be carried out after the survey was conducted. 35 villages in six talukas (equivalent to blocks/districts) were included in the sampling frame, the six talukas are located up to 150km apart from each other. GIZ collected the number of households in the 35 villages in 2016, on which basis a proportionate sampling was done aiming for a total sample size of 400. The sampling frame is different from Jharkhand and West Bengal in that one requirement was to create a sample frame of farmers with a group that will receive training after the survey (treatment group) and with a group of farmers that will not receive training after the survey (control group). Hence, the intended sample size was split by same parts to treatment and control group in each village. The total number of households sampled in each village is proportionate to the size of the village. The sample size was set after rounding to 432 households. The sampling of treatment households was randomly selected from the list of predefined farmers. The sampling of control households was achieved through the random walk. The exact sampling procedure of the random walk in West Bengal and Karnataka can be found in the Appendix on page 177. Figure A.2 on page 180 in the Appendix shows a map of the surveyed households in Karnataka.

Regarding the village and market survey, the same methods in all three regions were used: The village-level recall surveys were conducted by using Focus Group Discussions (FGDs). The sampling process of each FGD was to communicate with the village leader and to collaborate with him for identifying suitable participants. 10 to 15 people per group were requested to join the FGD. The only requirement was that age and gender were approximately equally distributed among the group participants.

The market surveys were included for collecting the prices of locally available products. Hence, inquiries were made per village regarding the next available market, and on the basis of all available markets in each study region, five to eight markets were randomly chosen for the market survey. Within the market, five to six market vendors were randomly chosen whose food prices were collected. Table A.1 on page 178 in the Appendix displays the numbers of the full sample.

2 Focus on India: Food Prices and Food Security

This chapter considers the macroeconomic environment of India and its influence on the individual food and nutrition security situation. We give an overview on recent developments and discuss historical trends.

2.1 Introduction

The Indian subcontinent has suffered many famines throughout its history. The last large-scale and widely devastating famine occurred 75 years ago in Bengal of modern-day East India and Bangladesh during which an estimated two to three million people died. The economic and political consequences were far-reaching and influenced an alternative way of thinking about famine prevention and explanation (see among others DeLong, 2003; Sen, 1981). The last famines occurred on a smaller scale in the regions of Bihar and Maharashtra in the 1960s and 1970s. At this time the development of modern agricultural technologies and improved seeds led to an increased food supply. Also known as the Green Revolution, it led to a particular increase in wheat production. Since the 1960s, cereal production tripled from less than 100 million tons to almost 300 million tons in 2016 (World Bank, 2017). In turn, caloric supply per capita increased tremendously and the growing Indian population could be nourished. Until the 1990s India has been an economy largely driven by the agricultural sector in terms of income generation. The Gross Domestic Product (GDP) grew in line with agricultural production and population growth. This meant on a year-by-year basis that GDP per capita was only slightly increasing until the early 1990s (World Bank, 2017). The early 1990s marked a turning point for the Indian economy as it experienced a large-scale economic crisis. Consequently, political reforms induced a financial and economic turnaround and eventually led to an exponential growth of GDP per capita that persists until today (DeLong, 2003). Figure 2.1 on the following page displays the described development.

2.2 Recent Economic Development

India's economy has been a success story since the reforms of the second half of the 1990s, with high and stable growth rates, a modernized and globally integrated economy and tangible improvements for the average citizen such as

Figure 2.1: Trends in economic growth, agricultural production and population growth in India, 1961-2016

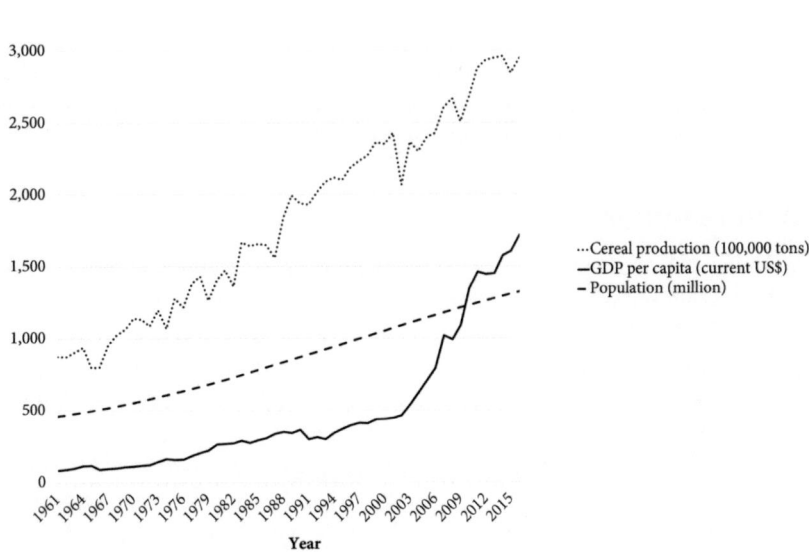

Data source: World Bank, 2017

a higher life expectancy and better education. More than 1.3 billion people live in India today, roughly a third in urban areas, with a recent population growth rate of 1.2% (United Nations, 2017). The median age is 27 years hinting at a huge and young labor force that can drive economic growth. The GDP per capita at purchasing power parity reached USD 7055 in 2017, up from around USD 1160 in 1991 (World Bank, 2017). The Gini coefficient as a measure of inequality was estimated to be 33.8 in 2013, indicating an unequal but not alarmingly unequal distribution of wealth (World Bank, 2013). These growth figures would have not been possible with a tightly regulated economic systems and a strong reliance on the agricultural sector as it had been before the 1990s[1]. The economic crisis in

---

[1] The economic system that was in place in India from 1947 to 1990 - called "License Raj" which is "the rule of the license" - had the character of a planned economy, including five year plans to centrally administer the economy by a Planning Commission and including trade limitations to follow an import substitution industrialization policy (DeLong, 2003).

the first half of the 1990s led to liberalization reforms aiming at a greater role of private and foreign investment through, among others means, the deregulation of markets, reduction of import tariffs, and reduction of taxes (DeLong, 2003). Disaggregating the growth figures that were induced by the reforms reveals the effects on the economy.

Estimates show that labor contributed only around 1% to the average GDP growth rate of 7% since 1990. Similarly, capital contributed only 1.5% to the average GDP growth rate. Instead, the economic growth in India is largely driven by productivity gains. Total Factor Productivity contributed more than 60% to the economic growth (Gupta et al., 2018). This productivity increase occurred primarily in the service sector, whereas in agriculture the gains were small and in the industrial sector the gains were more driven by employment increases (Bosworth and Collins, 2008). The World Bank estimates that future growth in India will be determined by accessibility to credits, savings rates and investment rates, which are all higher than in other emerging countries[2] (Gupta et al., 2018, p. 24).

The economic growth also trickles down to the poor. While in 1993/94 almost 45.3% of the population fell under the national poverty line, by 2011/12 that count was at 21.9% of the population being under the poverty line, which corresponds to receiving roughly USD 0.5 per day (Government of India, 2013). The average life expectancy at birth increased by an approximate 10 years since the early 1990s. The literacy rate rose from 52% in 1991 to 84% in 2011 (Government of India, 2018b). Yet, despite all these general positive prospects, there are regions and groups in India that are not benefiting from the economic boom. India is a patchwork of differences, and marginalized pockets exist particularly in rural areas.

The latest Demographic and Health Survey of 2015/16 in India exposed the rural urban divide regarding the nutrition and health situation (IIPS and ICF, 2017). For example, 26.7% of all women are underweight in rural areas in comparison to 15.5% in urban areas. At the same time, 31.1% of all women in urban areas are overweight in comparison to 15% in rural areas. In rural areas, 38% of children are stunted and 36% are underweight. Even more troublesome, 59.4% of all children and 52.4% of all women are anemic. Historical price and consumption trends help to explain this situation.

2.3 Food Price Trends

Prices reflect a twofold composition, a demand for goods as well as the supplied availability of these goods. In classical economic theory, consumer preferences are revealed through their demand for goods and the accordingly accepted prices

---

2 The emerging countries include China, Brazil, Mexico, Russia, Indonesia and Turkey.

Figure 2.2: Food price indices in India, three year averages from 1970 to 2017

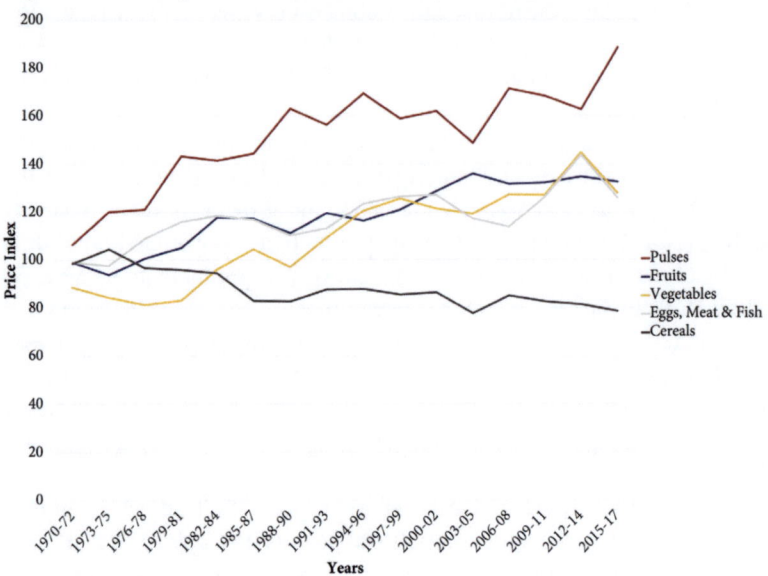

Data source: Government of India, 2018a. Wholesale prices with base year 1970 deflated with Consumer Price Index

(Houthakker, 1950; Samuelson, 1938). From a larger perspective, however, consumption preferences can also be induced by prices (von Weizsäcker, 2015). Considering the latter as well as the price development of food groups over-time, current incentives for a certain food consumption behavior can be unraveled (following Bouis et al., 2011). Figure 2.2 above shows the price indices of various food groups in three-year averages starting in 1970 and ending in 2017. The price indices of the past 50 years emphasize that cereals, the food group that generally supplies calories instead of micronutrients, have become relatively cheaper than all other food groups, including those which supply more vital micronutrients than calories. In other words, it is on average more costly to consume a balanced diet. From an individual point of view, these prices can be considered as exogenous trends; thus, these prices form an economic environment that incentivizes consumption of micronutrient-poor foods.[3]

---

[3] Similar trends have been described for other South Asian countries such as Bangladesh and the Philippines that also have high prevalence rates of micronutrient malnutrition (Bouis et al., 2011).

2.4 Food Consumption Trends

Considering the strong agricultural productivity and the price indices of various food groups for the past 50 years, it is not surprising that cereal prices decreased relative to the prices of other food groups such as fruits and vegetables. Developments in plant breeding, including the introduction of high-yielding crops during the Green Revolution, induced the high supply of cereals and the associated decline in prices (Bouis et al., 2011). This is a positive development from a food availability and accessibility point of view; however, this is also a disincentive for consuming diverse diets. To better understand the food availability of various food groups on a macroeconomic level, the food supply per capita can hint to actual consumption trends as Figure 2.3 shows below. Cereal consumption has been relatively stable for the past 40 years, indicating that the productivity gains met and partly surpassed demand. Despite price increases, consumption of legumes has been stable at a low level. Most notably, despite price increases of vegetables, fruits and dairy products, their consumption has almost doubled in the past 20

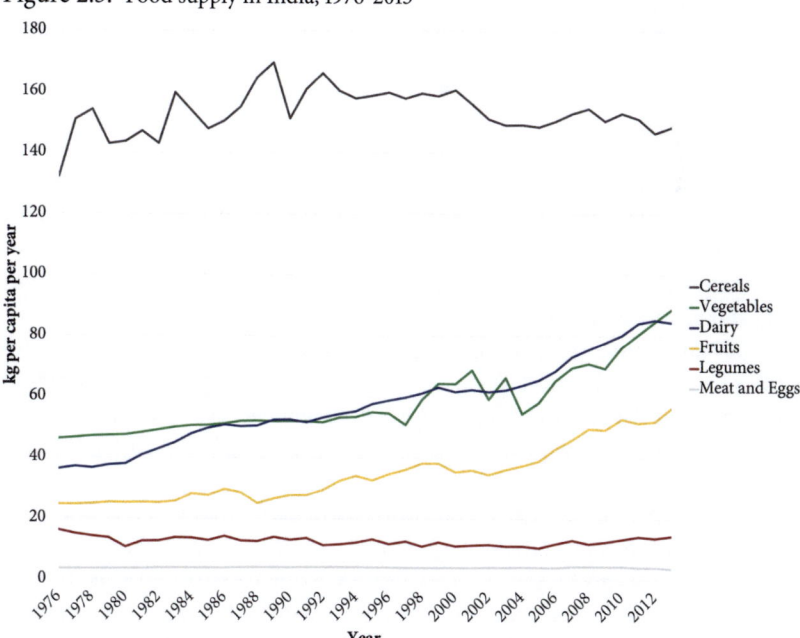

Figure 2.3: Food supply in India, 1976-2013

Data source: FAO, 2018b

Figure 2.4: Food supply of unhealthy foods in India, 1976-2013

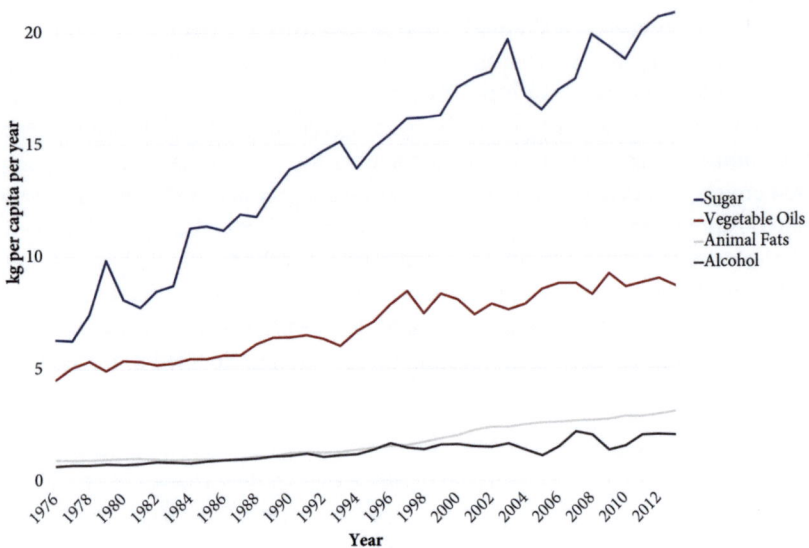

Data source: FAO, 2018b

years. The increased consumption of these food groups correlated with the increase of GDP per capita in India of Figure 2.1 on page 50. This possibly points towards increased demands of these foods due to a higher purchasing power, keeping in mind that per capita consumption cannot reveal any distribution of foods, nor any income-specific demands.

At the same time, the consumption of unhealthy foods such as sugar, vegetable oils, animal fats and alcohol has also increased, as the consumption trends in Figure 2.4 above display. This is a development that raises concerns for other malnutrition consequences, for instance obesity and non-communicable diseases.

Food and nutrition security is concerned particularly with vulnerable groups, hence with women of reproductive age and with children. Monitoring the overall trends in demand and supply of foods for these groups proves difficult. The Demographic and Health Surveys (DHS) that are conducted regularly in countries of concern try to fill this gap and provide nationally representative information on, among other things, nutrition intake and outcome. Based on the DHS data,

Figure 2.5: Food consumption by children in India, 2006-2015

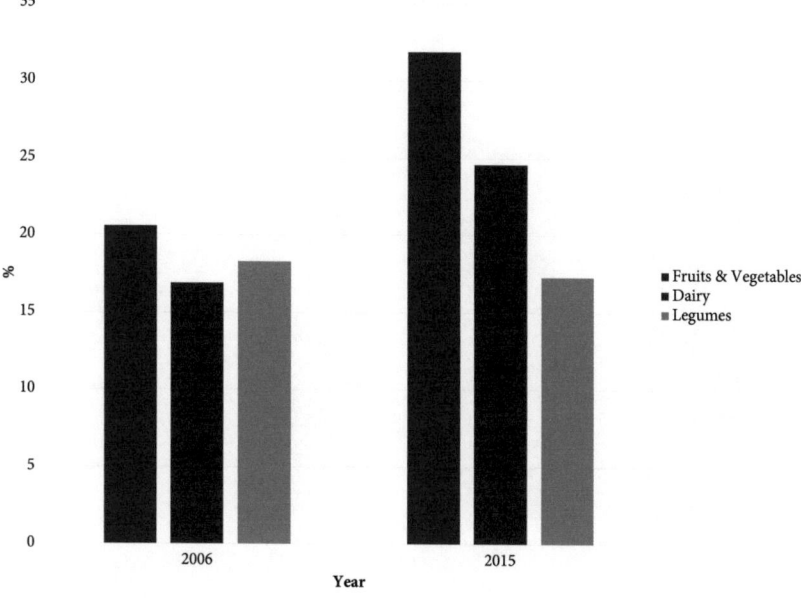

Date source: ICF, 2015

the growing consumption of fruits, vegetables and dairy products can also be recognized as a trend for children. Figure 2.5 above shows the share of children that consume certain food groups, indicating an increase of consumption in the past decade and stating a positive development on diverse consumption throughout age strata.

The UNICEF framework of nutrition prominently describes the link between nutrition intake and outcomes and clearly between income and nutrition outcomes (compare Section 1.3 on page 29). Indeed, in the past 20 years, during which food consumption and income per capita increased, the malnutrition indicators for children and women declined similarly as Figure 2.6 on the next page shows. As the dual burden of nutrition becomes prevalent, India currently has roughly as many underweight women as overweight women; at the same time, the undernourishment indicators of children are still high. A situation that was already hinted at by the food consumption trends. However, this double burden is split between rural and urban areas; being underweight is predominantly the problem in rural areas whereas being overweight is predominantly the problem of urban areas as can be seen Figure 2.7 on the next page.

Figure 2.6: Malnutrition rates in India, 2006-2015

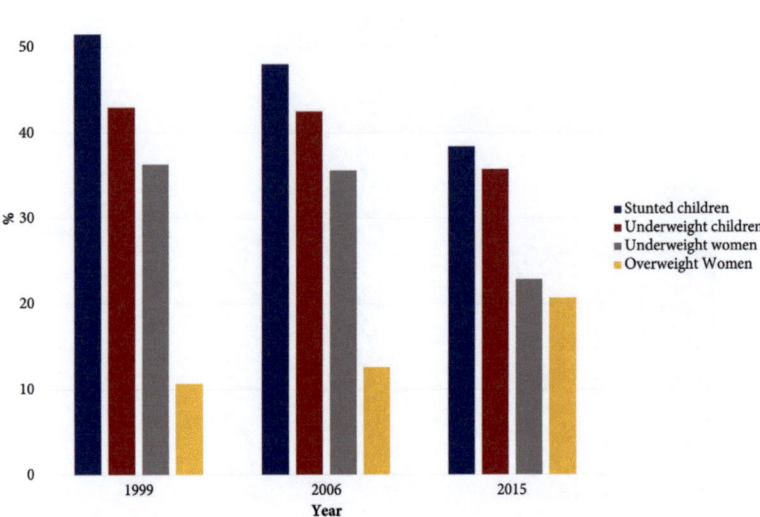

Data source: ICF, 2015

Figure 2.7: Malnutrition rates in rural and urban areas, 2015

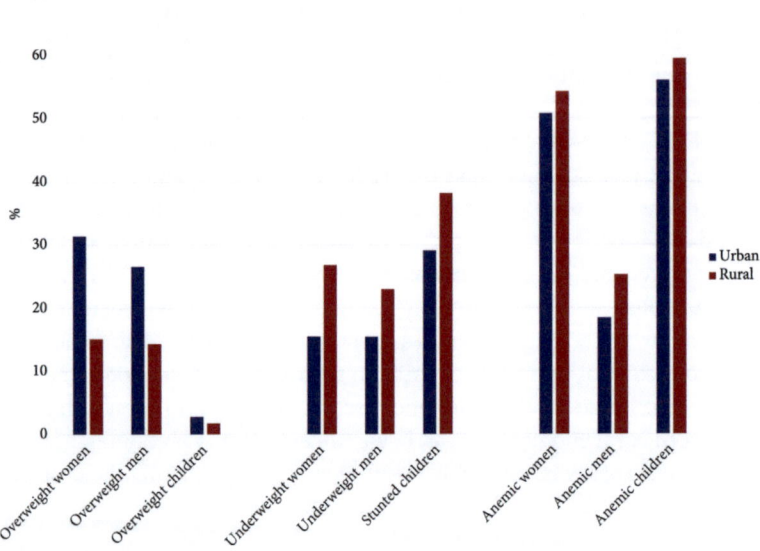

Data source: ICF, 2015

The descriptive numbers presented here draw a diverse picture of the food and nutrition security environment in India. Despite the increase in prices of micronutrient-rich foods, the consumption of these foods increased, presumably partly driven by income increases. Undernutrition rates decreased, while overnutrition rates increased. Increasing prices of certain foods raises the concern of preventing access to these for lower income groups. Indeed, the DHS data reveals that the average households with undernourished children is getting poorer[4]. Wealthier households can increasingly afford a diverse nutrition and benefit from better physical growth, whereas poorer households are continuing to suffer from malnutrition. Today, India faces food and nutrition security problems that are not as drastically acute as before the Green Revolution, but that are still concerning for individual and societal development.

---

4 The DHS wealth index categorizes a household's wealth in one of five categories. The average wealth index of households with at least one stunted child decreased from 2.7 to 2.4 between 2006 and 2015.

3 Production Diversity and Dietary Diversity

Abstract: On-farm production diversity of smallholder farmers can improve the nutrition security of the household. The objective is to determine the significance and relevance of this relationship by considering the different degrees of separability between both the commercial and consumptive production of food. A household-level survey is conducted covering socioeconomic, agricultural and nutritional data. The sample covers three regions of India including 811 households in 106 villages. Various regression specifications (OLS, Poisson, Probit, IV / non-IV) were used to estimate the effect of production diversity on dietary diversity. Variance of rainfall since 1997 is the excluded instrument. A positive association is estimated (β: 0.188 / 0.015 | IV / non-IV). Access to markets is found to be more relevant to improve dietary diversity. The increase is significant only for a few food groups (dairy products, nuts and vegetable), and it is the higher income groups that primarily benefit from market integration. In conclusion, production diversity does improve nutrition security, but the positive market effect is stronger for farming households that have a higher income.

Keywords: nutrition-sensitive agriculture, dietary diversity, agricultural production, markets

3.1 Introduction

Nutrition-sensitive agriculture is a buzzword that has attracted some attention in development implementation and scientific fields alike since the global food price crisis hit in 2007/08. Per Pinstrup-Andersen called for better research on how nutrition-sensitive food systems can effect nutrition security (2013) because implementation was limited to small-scale projects such as kitchen gardens up until then. The larger scale of nutrition-sensitive linkages can be understood as the effects of environmental biodiversity on human nutrition (Frison et al., 2011). More specifically, the diversity of agricultural food production systems can affect food consumption. Accordingly, this chapter looks at the allegedly non-nutritional factors of smallholder farmers' food production and the household members' dietary consumption.

There are various paths by which agricultural production can affect dietary intake of individuals and households alike. Prominently discussed is agriculture as a producer of food for farming households, agriculture as an income generator through which food can be purchased, and agriculture as a vehicle for decision-making power on intra-household food allocation through women's participation and empowerment (Ruel and Alderman, 2013). Hence, quality and quantity of dietary intake depend on the agricultural production. The linkages are particularly strong in rural areas where agricultural production takes place in smallholder settings. In these settings, the absolute yield is an approximation to the quantity of available food. Production diversity on the other hand can guide as a reflection of the diverse quality of production. Figure 3.1 below presents a conceptual framework that indicates the possible paths at the level of a farming household.

On-farm production is divided into commercial and consumptive production. The decision on the production share is made based on factors such as commodity prices, market integration, value chain development, and integration or storage capacities. Market-oriented production is primarily used for income generation, which creates the option for additional food purchases. The decision on the extent of food purchases is influenced by individual preferences, food availability or food prices. Food expenditures are also influenced by subsistence food production e.g. in case certain items are produced instead of purchased. The final quality and diversity of the dietary intake is a result of intra-household food

Figure 3.1: Conceptual framework for smallholder farming households

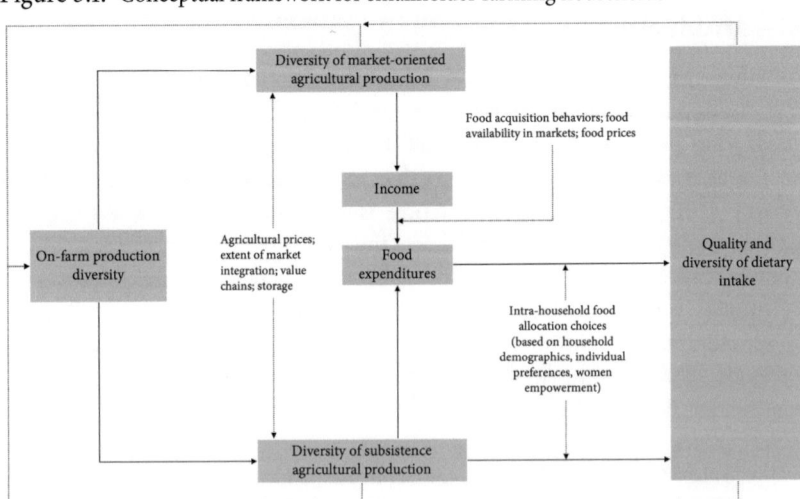

Author's illustration based on Jones (2017)

allocation choices. The future on-farm production diversity is again influenced by various factors; among these are consumption preferences and market demand considerations.

This chapter focuses on contributing to one research objective, namely, identifying the effect of production diversity on nutrition security. The two research questions are:

1. Do production choices affect nutrition choices of smallholder farmers and if so, to what extent?
2. Does market access influence this relationship?

3.2 Empirical Evidence and Research Questions

Links between production and consumption from a nutritional point of view have been quantitatively studied quite recently. Robust results were produced with nationally representative data from Malawi (Jones, Shrinivas, et al., 2014), in which the effects of production diversity was positively estimated to affect the HDDS in agricultural households. Production diversity was measured through the Simpson's Index, an index that considers the number of crops as well as the distribution area on which these are cultivated. The Simpson's Index also accounts for permanent crops and tree crops that grow in the designated area. HDDS and Simpson's Index are suboptimal measures. The HDDS does not measure dietary diversity, but the access to food (Swindale and Bilinsky, 2006). The Simpson's Index considers more crops than are considered in dietary diversity scores; a homogenization of measurements are necessary for a priori causality claims. Furthermore, given the non-separability condition of agricultural households, the linkage between agricultural production and consumption is further dependent on the access to markets (Singh et al., 1986).

In a cross-country comparison, Malawi, Kenya, Ethiopia and Indonesia were compared (Sibhatu et al., 2015c). Market access was interacted with production diversity to measure the combined effect on dietary intake. The authors found a positive effect, however market access seems to negate the effect of production diversity on consumption diversity. The methodological approach for measuring and comparing the questioned indicators was criticized as insufficient, giving rise to an interdisciplinary debate (Berti, 2015; Remans et al., 2015; Sibhatu et al., 2015b). The most recent economic approach to unravel the relationship contains a surprisingly similar caveat, though with different indicators (again the HDDS was used) (Koppmair, Kassie, et al., 2017). The academic debate recognizes the need for a more careful selection of indicators, but also to take into account the infrastructural aspects such as market access (Koppmair and Qaim, 2017a,b; Sibhatu and Qaim, 2018a; Verger et al., 2017). On the other hand, in a longitudinal set-up, Jones (2017) confirms the effect on dietary diversity but finds no negating effect of market access in Malawi.

While most of the research is done in Sub-Saharan Africa, there are only a few papers that take into account the South Asian setting. For example considering India, there is merely one paper and it does not show any correlation between production diversity and household dietary diversity (Kavitha et al., 2016). However, the paper measured production diversity using the Simpson's Index over the period of one year, instead of the last season. This is a questionable measure regarding perishable foods that are usually consumed shortly after harvest. Moreover, the survey was conducted in an area that primarily practices monocropping. In Nepal a positive relation was found (Malapit et al., 2015). However, this paper looked more into aspects of female empowerment and included production diversity as a control. A comprehensive meta-study comparing published studies on the relation of production diversity and diets adds further insight to the variability of indicators used and results estimated (Sibhatu and Qaim, 2018b). The authors find that the relevance and significance of the association is stronger in Sub-Saharan Africa than in South or Southeast Asia (Sibhatu and Qaim, 2018b, p. 15).

The presented literature is limited in aspects of causality. The causal linkages are primarily based on conceptual considerations. Dillon et al. (2015) tried to overcome this shortcoming by using an Instrument Variable (IV) approach, utilizing climate variability as an instrument for production. They conclude that climate variability could be a possible instrument, but their results indicate that climate variability is a better instrument for agricultural revenue in general, thus missing the point of reflecting production diversity. However, Dillon et al. present an identification strategy that presents weak instruments and needs to be read with caution (Dillon et al., 2015, p. 989). Hirvonen and Hoddinott (2016) propose temperature, altitude, the interaction of these and the slope of the farm land as an instrument. They find a positive effect on the diets of pre-school children in Ethiopia, and a particularly strong relation for households with partial access to markets. However, production diversity was calculated by considering the output of a full year, which is hard to justify for perishable goods in this particular context.

Accordingly, improvements to the research topic are proposed by the present study in the following way. (1) An appropriate methodology is used for measuring nutrition security and production diversity (see debates on Koppmair, Kassie, et al., 2017; Sibhatu et al., 2015c. (2) The identification strategy is clearly set out with one instrument (Dillon et al., 2015; Hirvonen, Hoddinott, et al., 2017; Sibhatu and Qaim, 2018b). (3) The robustness of results of existing literature can be improved and has to be considered (Hirvonen and Hoddinott, 2016, p. 11). (4) The effect of market access is clarified in this context. (5) The specific regional context is also important and as such, no information on South Asia is sufficiently provided (Jones, 2017, p. 94). Furthermore, this study looks particularly at smallholder households in India. Thereby, this study adds to the existing literature by using a homogenous data source from three different regions in India using cross-sectional surveys that were similarly conducted. The questionnaire design includes most of the indicators and information that were discussed by literature, which was previously not possible due to the recent developments in this field.

3.3 Theoretical Model

Although the effect of nutrition-sensitive agriculture and production diversity on the level of food and nutrition security is conceptually well outlined and globally applied by the development agenda, the theoretic foundation is not so widely discussed. The literature provides empirical exercises that mostly relate and sometimes causally determine the effects between nutrition-sensitive production choices and dietary intake. Methodological approaches are motivated from agricultural and nutrition sciences or from economics. But theoretical reasoning is notably lacking. One recent article (Allen and de Brauw, 2018) approaches this lack in the literature by describing a simple relation of consumption decisions that are subject to budget constraints for obtaining optimal diets. We consider Allen and de Brauw's discussion as a starting point from the perspective of economic theory. This section introduces a possible pathway to how a production decision might influence consumption decisions, hence, how production diversity might influence dietary diversity. The graphical visualization of this theoretical approach further explores why food insecurity situations might occur in household optimization settings.

Following Allen and de Brauw (2018), let us assume a simplified setting where the consumer can choose only between two food product categories. This setting regards a choice for budget allocation that is independent of any budget share for non-food expenditures. Instead we would like to explain a possible trade-off between micronutrient-dense and energy-dense foods. Therefore consider a staple food grain such as rice (q_g) and a mix of vegetables (q_n). The staple food q_g is energy-dense, whereas the mix of vegetables q_n is micronutrient-dense. The foods have associated prices (p_g and p_n) and the consumer has an available budget of m. We get the budget constraint:

$$p_g q_g + p_n q_n \leq m$$

In the following, we will consider the consumer to be a household that act as a homogeneous agent and that behaves in a consistently rational way for maximizing its utility[1]. The household has a utility function for the consumption of q_g and q_n, denoted as $U(q_g, q_n)$. The utility function is quasi-concave and can be differentiated twice with q_g and q_n. Without further specification of the utility function, we can describe a stylized convex indifference curve[2].

A household maximizes its utility subject to the available budget in order to receive an optimal consumption bundle of q_g^* and q_n^*. The optimal bundle can be derived with:

$$\frac{u(q_g)}{u(q_n)} = \frac{p_g}{p_n}$$

subject to

$$p_g q_g + p_n q_n = m$$

Indifference curve

The two graphs in Figure 3.2 on the following page show exemplary indifference curves of the above utility function. Figure 3.2a displays the available budget m and three possible indifference curves of which one is not dashed. In an optimal setting, the household would consume the available food bundle at point A_{pareto}, which we might denote as q_g^* and q_n^*. Point $A_{pareto}(q_n^*/q_g^*)$ reflects the pareto-optimal consumption given the budget m, or the point where the highest possible indifference curve (the not-dashed one) touches the budget line. If the household would consume any food bundle on the dashed indifference curve to the left, the household would consume a sub-optimal combination of foods; any food bundle on the dashed indifference curve to the right cannot be attained with the given budget.

---

1 We consider the household to live in a rural area that has an undefined number of members, but at least one couple.
2 We factor out all other, but probably equally important factors that might influence a household's preference of one food group over the other. These can relate to culture, personal preference, social interaction, emotions, demographics etc. These variables might influence the curvature of the utility function and is often depicted as Z, such that the utility function would be $U(q_g, q_n, Z)$.

Figure 3.2: Consumption indifference curves

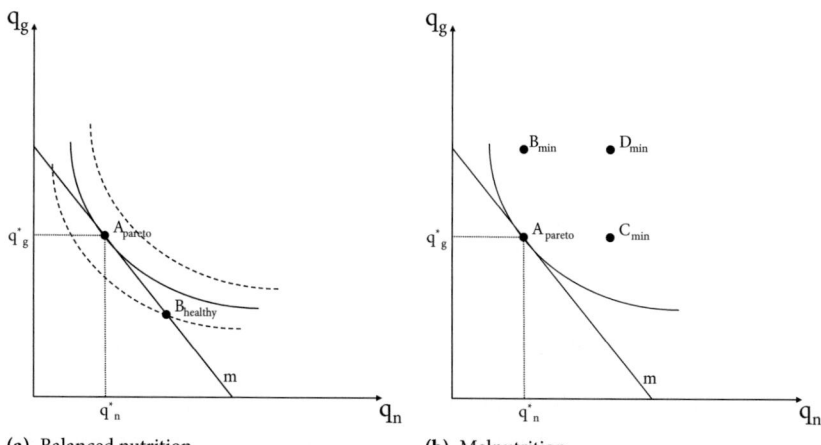

(a) Balanced nutrition

(b) Malnutrition

Figure 3.2a allows us to find an economic rationale for why households are not optimally nourished even if their available income would allow them to be. Assume that for a household the point $B_{healthy}$ indicates from a nutritional point of view the optimal combination of rice and vegetables. Since this point touches the budget line, it is still affordable for the household. However, given the utility specification, only an inferior indifference curve meets the point. In other words, the household receives a higher utility in malnourishment than in consuming a healthy set of foods and therefore will always chose the unhealthy option. There are two solutions to match a pareto-optimal food bundle with a healthy food bundle: either adjusting the utility function of the household or adjusting the prices of the foods. More generally speaking, either adjusting demand or supply. Chapter 4 and 5 will look at the preference side respectively how individuals form a certain demand for foods, whereas this chapter deals with one aspect of the supply side. However, we have yet to find the theoretical link between production and consumption.

This relatively simple framework also allows us to indicate the challenges of food and nutrition security as defined in Chapter 1.4. In the setting of this research, there are two problems that individuals are at risk of facing: the lack of calories and/or the lack of micronutrients. Figure 3.2b above displays the additional points B_{min}, C_{min} and D_{min}. If the household had more available budget, higher indifference curves could meet these points. The points reflect three different settings of minimum requirements regarding grain and vegetable consumption. B_{min} has the same q_n yet a higher q_g than $A_{pareto}(q_n^*/q_g^*)$, meaning the household consumption at A_{pareto} is lacking the minimum requirement of grain; therefore, the household is potentially short of calorie intake and faces undernutrition. C_{min} has a higher q_n yet the same q_g as $A_{pareto}(q_n^*/q_g^*)$, meaning the household consumption at A_{pareto} is lacking the minimum requirement of vegetables; therefore, the household is short of micronutrients and potentially faces hidden hunger issues. Finally D_{min} has a higher q_n and a higher q_g than $A_{pareto}(q_n^*/q_g^*)$, meaning the household consumption at A_{pareto} is lacking the minimum requirement of both grains and vegetables and the household is potentially in a severe situation of malnutrition. Hence, through the pareto-efficient allocation of foods, which a household will always chose to optimize its utility, various states of malnourishment can be explained.

Note three characteristics of both graphs in Figure 3.2, which hold true for the displayed graph as well as for the following discussion: *First*, we do not define the units of q_g and q_n, it is merely a hypothetical reflection of a preferred combination of foods that the household might demand. In the above equations, though, q_g and q_n reflect quantities. *Second*, the indifference curves never touch any axis. This reflects that a minimum amount of either q_g or q_n is needed to be consumed in order to meet the household's utility. However, the indifference curves are closer to the y-axis than to the x-axis, indicating that the household will always demand a higher amount of staple crops than vegetables. *Third*, the implicit prices are set such that staple crops are relatively cheaper than vegetables so that the slope of m is $\frac{p_g}{p_n} < -1$. This indicates the economic nature of vegetables as high value crops.

Production possibility frontier

When discussing the link between production diversity and dietary diversity, we necessarily need to analyze farming households; otherwise possible links would solely be based on markets. Figure 3.3 on the following page shows two graphs that link Production Possibility Frontiers (PPF) of a household with the possible consumption indifference curves of the same household. Figure 3.3a represents a purely subsistence farmer (which is more for theoretical considerations than for a realistic reflection) and Figure 3.3b represents a household that is integrated into a market system.

Looking at Figure 3.3a, an isolated farming household can only consume the food bundle that it produces. With the given resources, the household can produce any bundle of foods that is located on the curve PPF, assuming that a

Figure 3.3: Production possibility frontier and consumption indifference curves

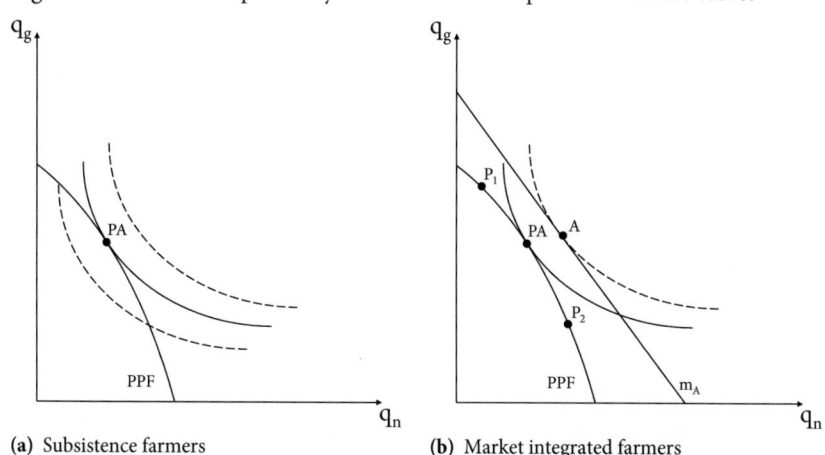

(a) Subsistence farmers (b) Market integrated farmers

household can produce with the same resources more grains than vegetables. Similarly to Figure 3.2a, we can see three indifference curves of which one touches the PPF at point *PA*. In an optimal case, the household will be able to produce exactly the bundle of foods that its a priori given utility function would require for meeting the pareto-optimum. In any case, the production will equal the consumption, or the available budget is in fact the PPF of the household.

Looking at Figure 3.3b, which reflects a more realistic setting, the available budget is more than the household could generate solely by its agricultural production. Market integration not only allows the household to sell its agricultural produce, but for example also to participate in the labor market or benefit from input markets; hence, the available budget for food purchases increases. It is crucial to state that a smallholder farming household is very unlikely to solely produce for markets and will very likely also consume part of its own produce. If purely market-oriented production would occur, the research question on how production diversity will affect dietary diversity of smallholder farmers would become irrelevant. In Figure 3.3b, m_A represents the available budget, and the pareto-optimal point *A* lies on an higher indifference curve than the previous, now inefficient consumption point *PA*. More precisely, we can see that the utility maximizing consumption includes a higher q_g as well as a higher q_n, thus dietary diversity can be increased. Accordingly, we can state a falsifiable hypothesis that can be deducted from this theoretical framework:

Market integration of households increases the consumption of foods and increases dietary diversity.

Transaction costs

Nutritious foods often face additional problems in comparison to grains. Looking at the value chains of the various food types, it is clear that nutritious foods such as fruits or vegetables, but also most of animal sourced products, require a quick transportation to the consumer. The transaction is time-sensitive, otherwise the produce might become unsafe or even decay. Preservation techniques such as appropriate drying or cold storages are expensive and are often lacking in marginalized rural areas. Moreover, these techniques can also reduce the nutritional content (Hodges et al., 2011). This additional burden can be economically described and modeled as transaction costs (Allen and de Brauw, 2018).

From a consumer's point of view, the resulting price for a good q_n is the sum of the producer's costs c_n and the transaction costs c_t (leaving aside any possible profit margin of the producer):

$$p_n = c_n + c_t$$

Respectively, for the available budget of a household, the affordable bundles can be calculated with:

$$p_g q_g + (c_n + c_t) q_n \leq m$$

Figure 3.4a on the following page displays the shift of the budget line due to the included transaction costs from m_A to m_B. The household is still active in agricultural production with production output at point P_1, and it is still integrated into the market. It is clear that due to the reduced purchasing power of the household in regard to q_n, only a lower utility can be met; hence, the pareto-optimum changes from A to B and a lower amount of vegetables will be consumed.

What happens if the household decides to use its available resources and to produce more diversely? Figure 3.4b indicates the shift of production from P_1 to P_2, which is still on the same PPF, so the household is neither intensifying nor extending its production realm. The household solely decides to produce more vegetables and less grain. The available budget will now shift from m_B to m_C. This shift can be explained in terms of transaction costs. In case the household consumes the same amount of vegetables as before the production shift, the household will pay less. This is because it saves the transaction costs that

Figure 3.4: Effect of transaction costs on consumption

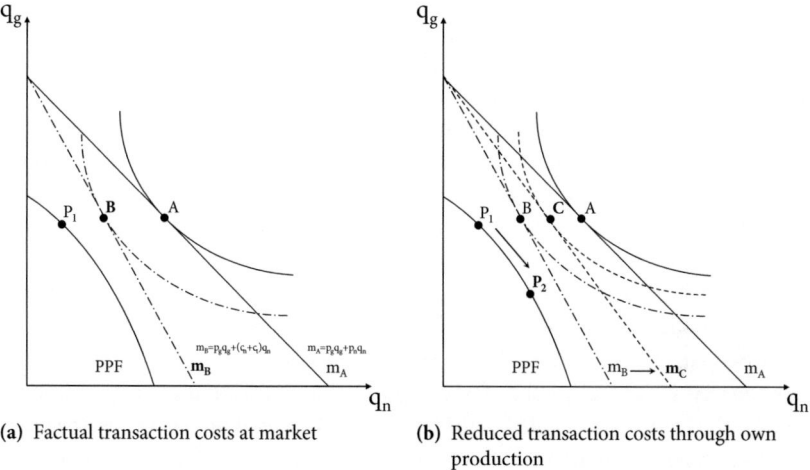

(a) Factual transaction costs at market

(b) Reduced transaction costs through own production

were previously imposed by other producers given that the household will consume at least part of its own produce. The transaction costs will not occur at all, since the produce can immediately be consumed after harvest. Hence, point C indicates that a higher amount of vegetables can be consumed and a higher pareto-optimum can be achieved. Yet, it is still inferior to the optimum that could be achieved without any transaction costs in point A. Obviously, the shift of curves is stylized in the graphs and does not represent any actual or possible shift in magnitude. The theoretical link between an increase in production diversity and dietary diversity can be explained with realized savings due to decreased transaction costs. Accordingly, we can frame a second hypothesis:

> *Increasing on-farm production diversity along the production possibility frontier increases the dietary diversity of the household and of individual household members.*

In the following sections we will empirically assess the two hypotheses and also find a causal statistical link between production diversity and dietary diversity.

3.4 Data

This study relies on primary data that was collected in the states of Jharkhand, West Bengal and Karnataka in India between January and June 2017. Village-level data and market information complements the household-level survey. The used data is a subset of a larger dataset that was collected for an overall research objective. The size of the initial dataset consisted of 1324 households in 119 villages of the 3 regions. The regions were chosen to suit the overall research objective (i.e. rural areas that have high prevalence rates of malnutrition and that are prone to environmental shocks such as droughts and floods). We utilized a two-stage cluster sampling technique: the villages were randomly chosen from pre-identified districts and the households were also randomly chosen, partly on the basis of a random draw from current census data. If the census data was not up to date, we used the random walk technique for identification of the respondent households (see on page 177 in the Appendix). The initial sample size was reduced to 811 households in 106 villages and 3 regions because of the focus on households with agricultural production (326 households/49 villages in Jharkhand, 64/22 in West Bengal, 423/35 in Karnataka). We have individual data for the household head as well as for the spouse of one's household. Nutrition information is available for the female household head/spouse and for at least one child below 2 years per household. The main variables to be used in this study are described in Table 3.1 on page 72.

The average household has 5.2 members, is male-headed and has a gender ratio of 0.48 women to men. The age of the household head is approximately 42 years, the age of the spouse is 33 years. The average household head has 6 years of formal education and 78% of the household heads are working primarily in agriculture-related activities either as a farmer or day laborer. However, all households are producing food products either through livestock or farming. The average income per capita is INR 1542 per month at adult equivalence (approximately USD 24)[3]. Table 3.2 on page 74 shows the summary statistics in detail.

3.4.1 Dependent variables

This study examines nutrition security as a set of various nutrition intake variables by estimating multiple regression models. Many various indicators exist to estimate the nutrition security of households and individuals. Concerning micronutrient deficiency, estimating the micronutrient content of human blood is the most exact approach, although also the most costly and time consuming. Proxy indicators such as the following are used instead (Martin-Prével et al.,

---

3   Income is calculated from expenditures and weighted based on an adult equivalence scale (AE) per household. The weighting is defined as follows, where age is in years: AE = 0.5 if age < 5, AE = 0.7 if $5 \leq$ age < 15, AE = 1 if age $\geq$ 15.

Table 3.1: Description of main variables

Variables	Description
Individual variables	
Dietary Diversity Score	Number of food groups consumed by woman in past 24 hours on a scale of 10 food groups
Age woman	Age of woman in years
Literacy of woman	Binary variable on literacy level of woman (1 = literate, 0 = illiterate)
Household variables	
Production diversity	Number of food groups produced by household in past season on a scale of 10 food groups
Age of household head	Age of household head in years
Highest formal education among adult household members	Most years of formal education among household members >14
Income	(log) Income of hh per adult equivalent per month based on expenditure
Religion	Categorical variable of household head's religion (1 = Muslim, 0 = Hindu)
Number of males 0-5 years	Count variable for number of males 0-5 years
Number of males 5-15 years	Count variable for number of males 5-15 years
Number of males 15-60 years	Count variable for number of males 15-60 years
Number of males 60+ years	Count variable for number of males 60+ years
Number of females 0-5 years	Count variable for number of females 0-5 years
Number of females 5-15 years	Count variable for number of females 5-15 years
Number of females 15-60 years	Count variable for number of females 15-60 years
Number of females 60+ years	Count variable for number of females 60+ years
Primary occupation of household is non-farm	Binary variable if primary occupation is non-farm (1 = yes, 0 = no)
Total land size	Total land available for agricultural production in hectares
Number of government schemes that were used by household	Count variable of number of government schemes that the household was utilizing (counted if at least one member is using a scheme)
Distance to next market	Distance to the next available market in km

Table 3.1: *(continued)*

Variables	Description
Household member visiting regularly next market	Binary variable if a member of the household is visiting the next available market regularly (1= yes, 0 = no)
Variance of rainfall	Variance of rainfall in mm² calculated for the past 20 years (1997-2016)
Village variables	
Village population	Population size of the household's village
No. of years that village is electrified	Number of years since the household's village received access to electricity
Regions	
Jharkhand	Binary variable if household lives in region Jharkhand (1 = yes, 0 = no)
West Bengal	Binary variable if household lives in region West Bengal (1 = yes, 0 = no)
Karnataka	Binary variable if household lives in region Karnataka (1 = yes, 0 = no)

2015). Food recalls for the past 24 hours, 7 days or even months are the most frequently used sources for proxy indicators. How to group the consumed food items and how to construct the indicators is the core debate in nutritional sciences about the measurement of micronutrient deficiencies. This study relies on the currently most investigated aggregation of food items, which presents the best possible proxy indicator[4]. The DDW is calculated as the number of food groups that have been consumed by women between 15 to 49 years from a list of 10 defined food groups in the past 24 hours. The MDDW is a dichotomous variable that takes the value 1 if at least 5 out of 10 defined food groups have been consumed by one individual in the past 24 hours (FAO and FHI 360, 2016). Whereas

---

[4] Note that this DDW is based on the same 10 food groups as the MDDW that is today the gold standard indicator for measuring dietary diversity.

Table 3.2: Summary statistics for variables

Variables	min	max	mean	sd
Individual variables				
Individual Dietary Diversity Score	1	7	3.63	1.20
Age in years	17	80	33.54	12.36
Literacy of female	0	1	0.54	0.50
Household variables				
Food groups produced on 10 food groups	1	8	3.05	1.17
Age of household head	20	90	42.34	13.66
Formal education of household head	0	17	5.95	4.30
Income	4.19	9.48	6.99	0.81
Hindu	0	1	0.91	0.29
Muslim	0	1	0.09	0.29
Number of males 0-5 years	0	3	0.45	0.67
Number of males 5-15 years	0	4	0.30	0.60
Number of males 15-60 years	0	5	1.67	0.87
Number of males 60+ years	0	2	0.22	0.42
Number of females 0-5 years	0	3	0.43	0.66
Number of females 5-15 years	0	4	0.37	0.70
Number of females 15-60 years	0	5	1.57	0.75
Number of females 60+ years	0	2	0.17	0.38
Primary occupation of household is nonfarm	0	1	0.12	0.33
Total land size	0.01	24.58	0.83	1.23
Number of government schemes that were used	0	7	3.09	1.42
Distance to next market (in km)	0.07	20.00	4.95	4.64
Household member visiting regional market regularly yes/no	0	1	0.75	0.43
Variance of rainfall (mm^2)	17606	47375	30132	6742
Village variables				
Village population	75	12000	1312.62	1483.76

Table 3.2: *(continued)*

Variables	min	max	mean	sd
Years that village is electrified	1	67	27.58	20.67
Regions				
Jharkhand	0	1	0.40	0.49
West Bengal	0	1	0.08	0.27
Karnataka	0	1	0.52	0.50

Sample size is 811 households

the DDW is a count variable that measures the extent of nutrition security, particularly for households in nutrition insecure settings where the higher the count, the better the nutrition is. The MDDW aims to reflect a minimum requirement for micronutrient adequacy.

$$DDW_i = n$$

$$MDDW_i = \begin{cases} 1 & \text{if } n \geq 5 \\ 0 & \text{if } n < 5 \end{cases}$$

The food groups used for the construction of the indicators are:

1. Grains, roots, tubers
2. Legumes
3. Nuts, seeds
4. Dairy products
5. Meat, poultry, fish
6. Eggs
7. Dark leafy green vegetables
8. Other Vitamin-A rich fruits and vegetables
9. Other not Vitamin-A rich vegetables
10. Other not Vitamin-A rich fruits

Figure 3.5 on the following page shows the distribution of the DDW within the sample. The MDDW is met by 28.8% of the sample. It is apparent that the majority of the population consumes 3 to 4 food groups per day, which is an insufficient amount from a micronutrient point of view. Even lower consumption is an indication for severe malnutrition.

Figure 3.5: Dietary diversity scores of women

Distribution of number of food groups consumed by women

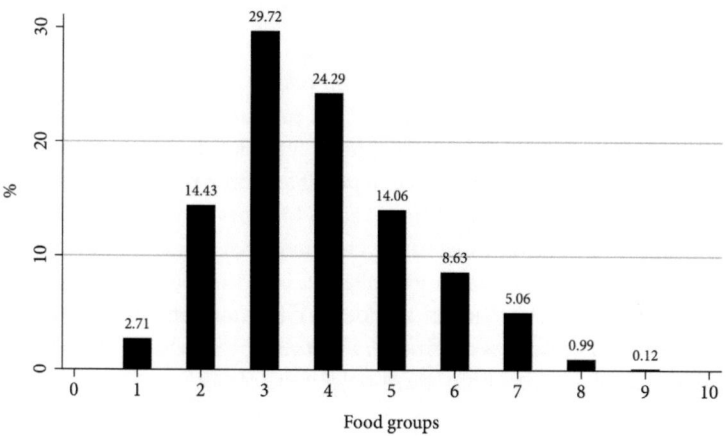

3.4.2 Main explanatory variables

Production diversity

The literature provides two distinct options to measure on-farm production diversity: either through a composite index or through a count of food groups. The composite indices reflect agrobiodiversity in a certain area. Frequently used are the Simpson Index and the Shannon Index (see Keylock, 2005). The Simpson Index reflects the inverse probability to find within a defined area the same plant species that covers a part of the area. The Simpson Index converges towards the maximum of 1 if an indefinite number of distinct species are habituated in the defined area. Similarly, the Shannon Index uses the distribution of area for distinct species, although it includes as well the distribution of area per species. The maximum value of the Shannon Index would be reached if all species cover the same area.

These indices are widely used for quantifying the biodiversity specific to various land uses in environmental systems. However, these lack the causal links to human nutrition diversity for the following reason. Looking at the example of a smallholder farm where wheat, barley and potatoes are grown, each crop being on the same size of area. This would reflect a low but existent diversity according to the Shannon and Simpson Indices. However, nutrition diversity would be 1 out of 10 (considering that nutrition security is represented by the number of food groups consumed on a scale of 10 different food groups as shown above for the dietary diversity scores). Moreover, livestock or poultry production can only be accounted for by modifying the indices, as no land area is attributed to livestock or poultry, which counteracts their intention. For instance, poultry might contribute to two food groups: eggs and meat. This increases their relevance for nutrition, but will not be reflected in the indices.

Therefore, we construct an indicator for production diversity that is constructed exactly as the dependent variable for nutrition security, i.e. the count of produced food groups on the basis of 10 possible food groups (Berti, 2015; Jones, 2017). The similar aggregation ensures causal linkages if relations exist. We will only consider crops that have been cultivated in the last agricultural season before the survey took place, for which we collected information on the species level. Dairy, meat and egg production is considered if livestock and poultry ownership (any compared to none) was confirmed. Figure 3.6 below shows the distribution of produced food groups within the sample of this study.

Market access

A market itself is here understood as a physical location at which food and non-food items are exchanged and on a once-weekly basis. Access to markets contains two different connotations: an infrastructural aspect (availability and accessibility) as well as the behavioral aspect (usage). Availability and accessibility is estimated by measuring the time it takes to reach the market starting at the household. We inquired the time in walking distance. Additionally, a household's

Figure 3.6: Production diversity of households

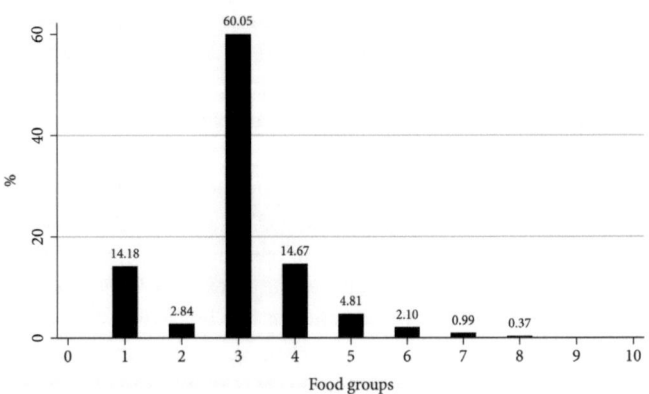

frequency of market usage is approximated by the inquiry of whether a member of the household is regularly visiting the next available market. The main explanatory variable is usage of the markets, whereas the distance to the next market serves as control.

3.5 Methods

3.5.1 Estimation strategy

The dependent variable DDW takes a non-negative integer value as these present count data: 0,1,2,.... From the descriptive statistics we can see that the dependent variables take only a few values with high probability (compare Figure 3.5 on page 75). Accordingly, we use a Poisson regression model for which we have the following basic specification:

$$Y_i = exp(\beta_0 + \beta_1 PD_h + \beta_2 X_{ihvr}) + \epsilon_{ihvr}, \qquad (3.1)$$

with r = 1,...,R, v = 1,...,V_r, h = 1,...,H_{vr} and i = 1,...,I_{hvr}. X_{ihvr} is a vector of explanatory variables and a set of control variables on the regional, village, household and individual level, β_2 is a vector with corresponding regression coefficients, and ϵ_{ihvr} is the error term. Of particular interest is the coefficient β_1 for PD_h representing production diversity on the household level.

As MDDW represents binary data taking only values of 0 or 1, we use a Probit regression estimation for the dichotomous variable as dependent variable, for which the basic equation is set as:

$$Pr(MDDW_i = 1 \mid x_{ihvr}) = \Theta(\beta_0 + \beta_1 PD_h + \beta_2 X_{ihvr} + \epsilon_{ihvr}), \qquad (3.2)$$

with r = 1,...,R, v = 1,...,V_r, h = 1,...,H_{vr} and i = 1,...,I_{hvr}. $MDDW_i$ can take 1 if the individual is consuming 5 or more food groups and 0 otherwise. The specifications are similar to the Poisson model: X_{ihvr} is a vector of explanatory variables and a set of control variables, β_2 is a vector with corresponding regression coefficients, and ϵ_{ihvr} is the error term. Of particular interest is the coefficient β_1 for PD_h representing production diversity on the household level.

In a second step, market access will be included in equation 3.1 as an interacting term, in which the coefficient β_2 represents the effect of PD_h interacted with MA_h and β_3 represents the effect of MA_h alone:

$$Y_i = exp(\beta_0 + \beta_1 PD_h + \beta_2 PD_h MA_h + \beta_3 MA_h + \beta_4 X_{ihvr}) + \epsilon_{ihvr} \qquad (3.3)$$

When estimating the main effect - the impact of production diversity on the various dietary indicators - we can assume to face various issues that limit the validity of the regressions. Given the cross-sectional data for the estimation, the following points should be of concern: (1) Ex ante it is likely that some of the control variables will affect production diversity such as assets. We assume further that perfect multicollinearity can be excluded. Yet multicollinearity will not reduce the predictability of our model, though we will not be able to claim with a sufficient validity the effect of the individual estimators such as the estimator of production diversity. (2) We are not able to completely comprehend all possible variables that affect diets. The claim that all variables are included in the model would be inadequate. Hence, it is possible to have unobserved characteristics that are correlated with either the dependent or the independent variables or with both, which creates the possibility for biased estimates. (3) Due to possible omitted variables, the model specification might also entail endogeneity issues, respectively the possibility that the main independent variables are correlated with the error term. Measurement errors can emphasize these issues. (4) Reverse causality could theoretically be of concern. Following the conceptual framework and the included time dynamics, a diverse diet might lead to an improved nutritional status, hence, to better suitability of the household for agricultural labor, which again might enhance the chances to diversify the household's agricultural production, particularly if market access is low (considering the non-separability condition). However, since we do not include any longitudinal data of the variables in question, we think the reverse causality issues can be safely ignored. Yet, the mentioned issues have to be addressed.

3.5.2 Identification strategy

In order to overcome the causal issues, we propose to use an instrument variable for production diversity. We know that 90% of all households have lived 22 years or longer in their respective villages (95% 12 years or longer, 50% 60 years or longer), hence, permanent migration is uncommon. Therefore, we can assume that differences in production diversity cannot necessarily be explained by migrated households with a preference for high production diversity (or low). Moreover, production choices depend among others on agro-ecological factors, which are locally-specific. Agro-ecological zones comprise geographical areas that are similar in production opportunities. Differences are induced by climatic conditions that are usually characterized through rainfall, temperature and elevation. Whereas the zones are large-scale, smaller clusters can also differ from surrounding zones in their characteristics. We propose to link the factors to local characteristics. Small-scale gridded data for rainfall, temperature and elevation is available either through the primary data collection or through secondary sources. Hirvonen and Hoddinott (2016) propose to use four instruments for an analysis in Ethiopia: temperature, altitude, their linkage and steep-sloping land. We propose to use the variation of rainfall as an instrument, which is sufficiently strong for representing an agro-ecological zone and for explaining agrobiodiversity differences in general.

For using the IV approach, the instrument needs to fulfill certain conditions (Wooldridge, 2013). Following equation (1), the proposed instrument variable (z_h) needs to be uncorrelated to the error term in (ϵ_{ihvr}) and sufficiently correlated to production diversity (PD_h):

$$Cov(z_h, \epsilon_{ihvr}) = 0$$

and

$$Cov(z_h, PD_h) \neq 0$$

If these conditions are met, we can claim that the instrument is exogenous in equation (3.1) and the instrument is relevant for explaining production diversity. Rainfall affects production choices in terms of food items (e.g. wheat needs less rainfall than rice), but in general there is no effect on food groups as these are defined for the dependent variables of dietary diversity (e.g. wheat and rice are both cereals). Accordingly, exogeneity can be claimed in regard to the outcome variable dietary diversity.

The mean of rainfall has been used as an instrument for income or economic growth (Brückner and Ciccone, 2011, prominently Miguel et al., 2004 and also in the Indian context Sarsons, 2015). Each application argues for its use in different ways. This study considers the variation of rainfall from an agricultural perspective in that we understand rainfall as reflecting agro-ecological zones, hence, reflecting agrobiodiversity on farmland in specific regions. Recent findings argue that the variation of rainfall has a positive effect on biodiversity; an effect that was found in various agroecological zones (Gherardi and Sala, 2015; Shimadzu et al., 2013; Yan et al., 2015). Accordingly, it can be claimed that the variance of rainfall is positively associated with agrobiodiversity, hence, possible production diversity.

On the contrary, one might argue that in a rural setting where the main source of income depends on agriculture, rainfall affects agricultural output, i.e. income. Income, on the other hand, has an effect on access to food and under certain conditions can enable households and individuals to consume more diverse foods (e.g. the likelihood of meat consumption increases). In settings with a minimal dietary diversity due to a missing purchasing power, this argument might hold true. Therefore, the exclusion restriction might be violated. This argument is often made considering the mean or absolute levels of rainfall, although never considering variation of rainfall.

Data for precipitation is taken from the Climate Hazard Group InfraRed Precipitation (CHIRPS) (Funk et al., 2015). It is a rainfall dataset that incorporates 0.05°(roughly 5km x 5km) resolution satellite imagery and balances it with station data. The grid is sufficiently small for observing heterogeneous rainfall at household level. The coordinates are matched with the household-level Global Positioning System (GPS) coordinates. Households were dropped from the sample, whose GPS coordinates' precision was larger than 1km. Daily data from 1.1.1981 to 31.12.2016 is available by CHIRPS. Various ways to measure rainfall were tested, but the strongest correlation and reliable F-tests between rainfall and production diversity in the first stage regression was found for the variance of rainfall from 1997 to 2016, in other words, over the past 20 years.

We use an Ordinary Least Squares (OLS) specification for estimating the correlation between the instrument and production diversity. The relevance is displayed by the Sanderson-Windmeijer F-test of excluded instruments (Sanderson and Windmeijer, 2016). On the basis of included covariates, which are the same as in the regression set up, variance of rainfall since 1997 is highly significant and positively correlated with production diversity. The F-statistic of 13.66 shows that the instrument is sufficiently strong considering the cut-off value of 10 (compare Staiger and Stock, 1997). Table B.1 on page 182 in the Appendix shows the results of the first stage estimation. We include the instrument in the second stage regressions, for which the Generalized Method of Moments (GMM) for the Poisson IV and Probit IV regressions is used and the Two-Stage Least Square (2SLS) for the Linear IV[5].

To exclude the statistical possibility that the chosen instrument variable might affect other relevant variables (e.g. dietary diversity, income or non-farm employment), we regressed the variance of rainfall on these variables including all covariates that are included in the final specification. The results show no significant correlation between rainfall variation and the variables in question (see Table B.2 on page 183 in the Appendix). Furthermore, testing for significant correlations between the residuals of the first stage regression and the dietary diversity

---

5 For an exactly identified model, which is the case here, the 2SLS estimator is the efficient GMM estimator and for its ease of implementation the preferred estimator for Linear IV (Hayashi, 2000, pp. 226-227). For Poisson IV and Probit IV we are limited to the GMM estimator, hence, its use.

with included instruments (as according to Hausman, 1978) yields a high significance at .99 level, thus indicating that endogeneity is indeed a problem in this model specification and adding further strength to the conclusion that the applied identification approach is necessary.

3.6 Results

3.6.1 Primary results: Effects of production diversity

Dietary diversity of women

Table 3.3 on the following page shows the results of the regression, in which Dietary Diversity of Women is the dependent variable and Production Diversity the main explanatory variable with controls on individual, household and regional level[6]. Focusing on the first two columns (1) OLS and (2) Poisson, PD is significantly correlated at the 0.1 level with DDW in the main Poisson specification, although the effect is quite small indicating a 1.5% increase of DDW per additional food group produced on the farm[7]. The OLS specification does not confirm these results as these are not significant (using the log of DDW for comparative purposes with Poisson). However, the β value of PD has a similar magnitude.

In comparison with other studies, we can see that the effect sizes are in an expected range. An estimated coefficient for production diversity pooled over four countries (Indonesia, Kenya, Ethiopia and Malawi) reflects a 0.9% increase of dietary diversity ranging from 5.4% in Indonesia to 0.2% in Ethiopia[8] (Sibhatu et al., 2015a). However, the effect size is on the low end. For example in Malawi, agricultural biodiversity is associated with an increase in dietary diversity ranging from 8% to 13% (Jones, 2017). Comparing the results with a recent review of available studies that analyzed the linkage between production diversity and nutrition (Sibhatu and Qaim, 2018b) is additionally helpful. For Asian countries the effect tends to be weakly significant and the effect size between production diversity and nutrition intake seems to be smaller than in other parts of the world; the

---

6 The relationship between DDW and Production Diversity (PD) has been tested for a non-linear relationship. Including a squared term for PD does not result in a better fit considering the R^2, neither does a link test indicate a specification error.

7 For the interpretation of the Poisson coefficients, we calculate the Incidence Rate Ratio (IRR) of each coefficient, which is $exp(\beta)$. $exp(0.015)$ results to 1.015, hence, to 1.5% increase.

8 The effect sizes seem to be coinciding, yet the previously mentioned flaws in the methodology of calculating production diversity and dietary diversity need to be kept in mind (see Section 3.4).

Table 3.3: Impact of production diversity on dietary diversity of women

	(1) OLS	(2) Poisson	(3) Linear IV (2SLS)	(4) Poisson IV (GMM)
Dependent variable	(log) DDW	DDW	(log) DDW	DDW
Production Diversity	0.016	0.015*	0.141	0.188*
	(0.010)	(0.009)	(0.093)	(0.106)
Individual variables				
Age of woman	−0.013	0.018	−0.098	−0.073
	(0.065)	(0.058)	(0.088)	(0.086)
Literacy of woman	0.080***	0.056**	0.059	0.040
	(0.030)	(0.027)	(0.037)	(0.034)
Household variables				
Age of household head	0.003**	0.002*	0.003**	0.002
	(0.001)	(0.001)	(0.001)	(0.001)
Highest formal education	0.004	0.003	0.005	0.004
	(0.005)	(0.004)	(0.005)	(0.005)
(log) Income	0.097***	0.097***	0.093***	0.085***
	(0.022)	(0.019)	(0.025)	(0.024)
Religion (0 = Hindu, 1 = Muslim)	−0.008	−0.024	0.022	0.015
	(0.057)	(0.056)	(0.071)	(0.075)
Number of males 0-5 years	0.007	0.008	−0.017	−0.018
	(0.024)	(0.022)	(0.030)	(0.029)

Table 3.3: (continued)

Dependent variable	(1) OLS (log) DDW	(2) Poisson DDW	(3) Linear IV (2SLS) (log) DDW	(4) Poisson IV (GMM) DDW
Number of males 5-15 years	0.021 (0.020)	0.014 (0.017)	0.005 (0.024)	-0.010 (0.028)
Number of males 15-60 years	-0.024 (0.018)	-0.019 (0.016)	-0.035 (0.021)	-0.030 (0.021)
Number of males 60+ years	-0.004 (0.036)	-0.010 (0.031)	-0.007 (0.041)	-0.019 (0.039)
Number of females 0-5 years	-0.003 (0.024)	0.009 (0.022)	-0.021 (0.030)	-0.011 (0.031)
Number of females 5-15 years	-0.003 (0.019)	-0.002 (0.016)	-0.014 (0.024)	-0.020 (0.025)
Number of females 15-60 years	0.025 (0.019)	0.020 (0.017)	0.024 (0.022)	0.019 (0.021)
Number of females 60+ years	-0.037 (0.041)	-0.014 (0.035)	-0.040 (0.045)	-0.022 (0.042)
Nonfarm occupation	-0.009 (0.043)	0.013 (0.040)	-0.027 (0.048)	-0.016 (0.052)

Table 3.3: (continued)

Dependent variable	(1) OLS (log) DDW	(2) Poisson DDW	(3) Linear IV (2SLS) (log) DDW	(4) Poisson IV (GMM) DDW
(log) Total land size	0.029*	0.024	0.005	−0.015
	(0.016)	(0.015)	(0.028)	(0.036)
Distance to next market	0.007*	0.004	0.010**	0.006
	(0.004)	(0.003)	(0.005)	(0.004)
Regular market visit	0.010	0.031	0.006	0.030
	(0.028)	(0.026)	(0.031)	(0.032)
Government schemes	0.022**	0.024***	0.031***	0.035***
	(0.009)	(0.009)	(0.011)	(0.012)
Village variables				
(log) Village population	−0.006	−0.025	0.033	0.025
	(0.020)	(0.017)	(0.032)	(0.034)
Years that village is electrified	0.002	0.002	0.002	0.002
	(0.002)	(0.001)	(0.002)	(0.002)
Regions				
West Bengal	0.307***	0.374***	0.216**	0.251**
	(0.076)	(0.072)	(0.094)	(0.106)

Table 3.3: (*continued*)

Dependent variable	(1) OLS (log) DDW	(2) Poisson DDW	(3) Linear IV (2SLS) (log) DDW	(4) Poisson IV (GMM) DDW
Karnataka	0.160*	0.228***	0.192*	0.302***
	(0.088)	(0.079)	(0.099)	(0.107)
Observations	811	811	761	761
Adjusted R^2	0.291		0.183	
Pearson goodness-of-fit		289.69		
p-value		1.00		
Sanderson-Windmeijer F-test			13.66	
p-value			0.0002	

Robust standard errors clustered by household in parentheses
*** p<0.01, ** p<0.05, * p<0.1

literature finds a mean positive effect of 5.6% (compared to 8% in Sub-Saharan African countries)[9].

Using the instrumented approach in columns (3) and (4) of Table 3.3, the results of the Poisson IV (GMM) regression show still a significant effect of PD on DDW at the 0.1 level but with a higher relevance, indicating an increase of DDW by 20.7% per additional food group produced[10]. Notably, the Linear IV 2SLS with log DDW is not significant and therefore cannot confirm the results.

The increase in the effect size might seem surprising at first sight; however, conceptually an increase in the effect size can be expected when using the instrument approach. Furthermore, looking at other studies that included an IV approach can bring the results into perspective. Hirvonen and Hoddinott (2016) estimated the effect of production diversity on child dietary diversity, finding a positive effect of 9.2% for a Poisson specification and 49% for a Poisson IV (GMM) specification. The IV result of Dillon et al. (2015) can be compared in effect size (although not in regard to interpretation due to weak instruments). The positive effect increases to 24% up from 3.7% in comparison to OLS results, however, also reflecting weakly significant results at 0.1 level. Thus, the primary results in Table 4 considering IV and non-IV estimates seem to be in line with previous findings.

Significant covariates vary between Poisson and OLS equations. Focusing only on those covariates that have a significant level of 0.05 or 0.01 or that have robust results throughout the various specifications, literacy of the women – a dummy variable if the woman for whom the dietary diversity score is generated can read and write or not – is positively correlated to DDW as expected. Income has a very robust and significant effect on dietary diversity. These results show that income is a highly decisive factor for increasing nutrition, though the channel is not yet clarified. Conceptually, income can only increase nutrition if the purchase of foods is possible. The next step of this analysis will shed more light on this.

The number of government schemes is similarly significant and robust. This is an indication that supportive policies, which are in place for poor populations, are effective. We do not disentangle the various programs and their specific effectiveness as it is not the focus of the research objective, nevertheless the frequency of usage can be reported. The most frequently used programs are the Public

---

9 Considering 40 studies of which 75% find mixed or positive results and of which only 3 are in the Asian context (Sibhatu and Qaim, 2018b, pp. 3-11)
10 Calculating the IRR: $exp(0.188)$ results to 1.207, hence, to 20.7% increase.

Distribution System (PDS) (by 95% of all households of which 80% are eligible for Below Poverty Line cards)[11], Anganwadi centers (65%)[12], and the National Rural Employment Guarantee (NREG) (54%)[13]. Given the widespread participation at these programs, it is likely that they are the drivers behind the positive effect[14].

Additionally, age of the household head and the regional dummies affect nutrition. The age of the household head can merely indicate a correlation. A slight positive effect could be related to a general tendency of the predominantly male household head being more altruistic towards his spouse and providing her with additional food when he is older (Chapter 5 will empirically assess this altruistic behavior). It could also relate to a stronger economic basis of older household heads due to possible remittances. But these are speculations and this study cannot provide sufficient evidence at this point. The regional dummies reflect the general statistics, which were the basis of study site selection (see Section 1.6.1 on page 42 about the research design). Jharkhand is the included region of the presented results and represents the poorest region with the worst nutritional indicators. In general, West Bengal and Karnataka are better off, which is displayed in the results.

---

11 The PDS distributes subsidized foods and non-food items to eligible households. The eligibility is determined upon income levels and categorized in three levels of income: Above Poverty Line (eligible for 15kg of food grains per month), Below Poverty Line (25kg to 35kg), and Antyodaya (35kg).

12 Anganwadi centers are rural child care centers that aim to improve child nutrition by food rationing, education and support in vaccinations.

13 NREG guarantees 100 days of wage employment per year to volunteers who do unskilled manual labor.

14 The usage of the policy schemes has been surveyed in a bivariate form, whether a member of the household the specific scheme or not. Nevertheless, testing the individual policy schemes with the given data for their effects on dietary diversity results in a non-robust picture. Only NREG produces significant and robust results in a regression that uses the same covariates as the previous analysis does (see Table B.3 on page 184 of the Appendix). It can only be hypothesized that the positive effect of NREG on DDW is mediated through the household income. Other policy schemes might still have a significant effect on DDW, but with the available data a thorough testing is not possible and rather opens up the opportunity for further research.

Minimum Dietary Diversity of Women

Table 3.4 on the following page shows the results of the regression, in which the Minimum Dietary Diversity-Women is the dependent variable and Production Diversity the main explanatory variable with controls on the individual, household and regional level. First of all, Probit and Probit IV (GMM) indicate that an increase in PD increases the probability that the minimum dietary diversity is met at the 0.05 and 0.01 significance level respectively. Hence, these results are more significant than PD has on dietary diversity when DDW is measured as a continuous count variable. The relevance of PD is increased in the Probit IV (GMM) estimation, increasing the marginal effects of production diversity from 3.6% to 63.8%. Hence, the probability of meeting the minimum adequate diet rises by 63.8% if the production diversity changes by an infinitesimal amount considering the presented model with all covariates. Looking again first at the Probit estimation, the age of the woman, number of females below 5 years in the household, government support and visits to the markets are increasing the probability and are significant. A higher number of adult males in the household (15 years and above) reduces the probability. The Probit IV (GMM) restates the positive effect of government schemes, but it also indicates the positive effects of regular market visits. The latter will be the focus of the secondary results in the following section.

3.6.2 Secondary results: Effects of markets

The primary results indicated that markets can have a positive effect on dietary diversity. To estimate the effect more precisely, we include market access as interaction with production diversity in the regression. Results for OLS, Poisson, Linear IV (2SLS) and Poisson IV (GMM) for estimating the market effect are presented in Table 3.5 on page 92. All specifications use the same covariates, although only the relevant interaction terms are presented. For the detailed table including the estimation results of the covariate see Table B.4 on page 189 in the Appendix. The approach for the IV estimations is that PD and the interaction term PDxMarket_visit are instrumented by Rainfall and RainfallxMarket_visit, thus the model is exactly identified. The Sanderson-Windmeijer F-test as calculated for the Linear IV (2SLS) specification indicate that the instruments are sufficiently strong with a F-statistic of 17.98 and 10.66 respectively (Staiger and Stock, 1997). We can assume that the same holds true for the Poisson IV (GMM) specification. The main focus of the discussion is placed on the Poisson and Poisson IV (GMM) specifications as the theoretically appropriate models given dependent variable being a count variable.

Table 3.4: Impact of production diversity on minimum dietary diversity of women

	(1) Probit Mfx / SE	(2) Probit IV (GMM) Mfx / SE
Dependent variable	MDDW	MDDW
Production Diversity	0.036**	0.638***
	(0.053)	(0.180)
Individual variables		
Age of woman	0.117	−0.079
	(0.300)	(0.312)
Literacy of woman	0.026	−0.041
	(0.137)	(0.135)
Household variables		
Age of household head	0.001	0.005
	(0.006)	(0.006)
Highest formal education	0.001	0.016
	(0.021)	(0.020)
(log) Income	0.101***	0.240**
	(0.091)	(0.109)
Religion (0 = Hindu, 1 = Muslim)	−0.134	−0.168
	(0.370)	(0.328)
Number of males 0-5 years	0.014	−0.054
	(0.140)	(0.136)
Number of males 5-15 years	0.019	−0.029
	(0.098)	(0.097)
Number of males 15-60 years	−0.039*	−0.165**
	(0.079)	(0.077)
Number of males 60+ years	−0.067*	−0.266**
	(0.147)	(0.135)
Number of females 0-5 years	0.072**	0.193
	(0.121)	(0.141)
Number of females 5-15 years	0.005	0.008
	(0.088)	(0.088)
Number of females 15-60 years	0.034	0.071
	(0.089)	(0.084)
Number of females 60+ years	0.025	0.056

Table 3.4: *(continued)*

	(1) Probit Mfx / SE	(2) Probit IV (GMM) Mfx / SE
Dependent variable	MDDW	MDDW
	(0.155)	(0.143)
Nonfarm occupation	0.081	0.099
	(0.246)	(0.237)
(log) Total land size	0.032	−0.014
	(0.083)	(0.101)
Distance to next market	0.005	0.022
	(0.015)	(0.014)
Regular market visit	0.112***	0.339**
	(0.143)	(0.159)
Government schemes	0.032**	0.151***
	(0.049)	(0.050)
Village variables		
(log) Village population	−0.016	0.123
	(0.088)	(0.102)
Years that village is electrified	0.001	0.006
Region fixed effects	Yes	Yes
Number of observations	811	761
Pseudo R^2	0.314	
LR chi2	203.381	350.070
Prob > chi2	0.000	0.000
Baseline predicted probability	0.289	0.292

Marginal effects at means. Robust standard errors clustered by household in parentheses
*** $p<0.01$, ** $p<0.05$, * $p<0.1$

First of all, we can recognize that PD (A), PDxMarket_visit (B) and Market_visit (C) have a similar direction of effects independent of the model specification. All effects are highly significant at significance levels of 0.05 and 0.01. The effect sizes of (A), (B), and (C) are relative to each other within each model specification of similar magnitude. We see that the interaction has a relatively small but still positive effect on DDW if the household is not regularly visiting the next available market. Considering Poisson IV (GMM), the increase of PD by one food group will result in an approximate 47% increase of the DDW[15]. That is a larger result than in the Poisson IV (GMM) specification of Table 3.3 but still within the expected range. If a household member is visiting the next available market regularly, the effect size of the interaction is an increase of approximately 3.38 times the DDW[16]. This large effect-size is caused by the model fitting and has to be interpreted carefully. It exemplifies much more the endogenous nature of the primary estimator PD[17]. Given the similar and robust results of all specifications, we can infer that PD has a higher singular positive effect on DDW if markets are not visited keeping all other things equal. Moreover, we underestimate the effect size of production diversity if we do not treat the endogeneity (as seen for child dietary diversity, e.g. at Hirvonen and Hoddinott, 2016, p. 8).

The result that market visits have a positive effect on DDW considering interaction with PD is presented visually in Figure 3.7 on page 93. This figure shows the effect of PD on DDW for each possible production choice of the farming households. Two trend lines are displayed, one without regular market visits ("w/o") and one with regular market visits ("with"). The shaded areas represent the 95% confidence interval for each estimated point. The increasing lines reflect the overall positive effect of PD keeping all other things equal. With market access, the line is higher overall, showing the positive effect of market access. We can further see that the difference between both lines is significant whenever the confidence interval is not touching the other line's points, which is the case for each PD level besides PD greater than six. The confidence intervals increases due to the decreasing number of households that produce the high number of food groups (compare Table 3.7 on page 76).

After inferring that market access in interaction with production diversity has a positive and significant effect on dietary diversity, it is useful to understand in which food groups this effect is relevant. Table 3.6 on page 94 shows the various food groups that form dietary diversity and production diversity respectively

---

15 Calculate: (0.386) x 1 + (-0.361) x 0 + (1.193) x 0 = 0.386 resulting in an IRR of $exp(0.386)$, which is a 0.47 times higher DDW considering the Poisson IV (GMM) model.

16 Calculate: (0.386) x 1 + (-0.361) x 1 + (1.193) x 1 = 1.218 resulting in an IRR of $exp(1.218)$, which is a 3.38 times higher DDW considering the Poisson IV (GMM) model.

17 The fitted model estimates the DDW for individuals without market access to be 1.7 and for individuals with market access to be 5.7, hence, the increase by approximately 3.4 times when including the effect of PD: (5.7/1.7). This does not reflect true values but the model estimate, therefore also consider the confidence interval of the estimates. Figure B.1 on page 185 in the Appendix visualizes this relation.

Table 3.5: Impact of market access on dietary diversity of women

	(1) OLS	(2) Poisson	(3) Linear IV (2SLS)	(4) Poisson IV (GMM)
Dependent variable	(log)DDW	DDW	(log)DDW	DDW
PD (A)	0.047*** (0.017)	0.045*** (0.016)	0.413** (0.186)	0.386*** (0.114)
PD X Market visit (B)	−0.045** (0.017)	−0.042** (0.016)	−0.438** (0.186)	−0.361** (0.114)
Market visit (C)	0.147** (0.066)	0.159*** (0.062)	1.323** (0.654)	1.193** (0.464)
Individual covariates:	Yes	Yes	Yes	Yes
Household covariates:	Yes	Yes	Yes	Yes
Village covariates:	Yes	Yes	Yes	Yes
Region fixed effects:	Yes	Yes	Yes	Yes
Number of observations	807	807	756	756
Adjusted R-squared	0.288			
Pearson goodness-of-fit test		288.0		
p-value		1.00		
Sanderson-Windmeijer F-test (A)			17.98	
p-value			0.000	
Sanderson-Windmeijer F-test (B)			10.66	
p-value			0.001	

Robust standard errors clustered by household in parentheses
*** p<0.01, ** p<0.05, * p<0.1

Figure 3.7: Predictive margins of market visits

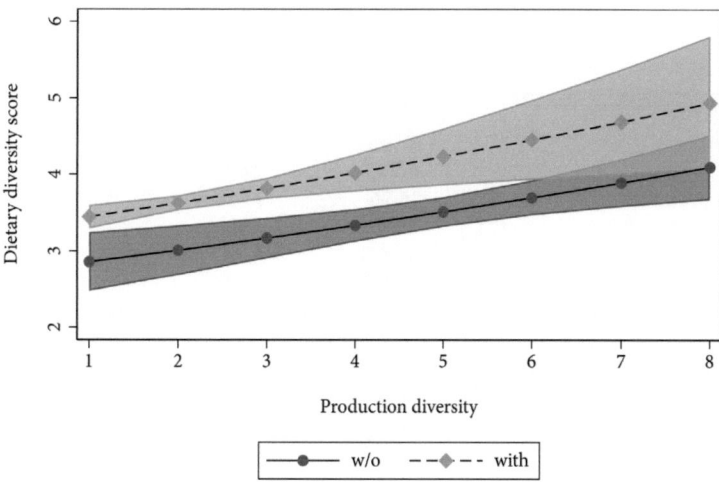

(compare with Section 3.4.1). This table shows the percentage of the sampled households whose women have consumed the respective food group. The χ^2 p-value indicates if the difference between being with and without market visits is significant[18]. The food groups "Nuts, seeds", "Dairy products", and "Other (not Vitamin-A rich) vegetables" are the food groups that are more frequently consumed if households visit markets, indicating that these food products are more likely purchased than produced on-farm[19].

As the next step, we want to understand if the consumption increase of the food groups is distributed equally over the population or if certain groups benefit more from market access and the corresponding consumption increases. In the following, three figures are presented displaying the change in consumption

---

18 The significance has been also estimated by using Probit regressions, having a food group as dependent variable and using all control variables as in the previous regressions of this study. Therefore it is sufficient to indicate only the χ^2 p-values at this point.

19 Regression results between market visits and the probability of food group consumption that confirm the presented χ^2 p-values of Table 3.6 can be found in Table B.5 on page 190 in the Appendix.

Table 3.6: Consumption of food groups by market visits

Food group	Consumption of food groups by regular market visits (in %)		χ^2 p-value
	no	yes	
Grains, roots, tubers	99.51	99.67	0.737
Legumes	68.47	70.16	0.650
Nuts, seeds	**6.90**	**14.92**	**0.003**
Dairy products	**16.26**	**33.11**	**0.000**
Meat, poultry, fish	8.87	5.90	0.142
Eggs	3.45	2.79	0.630
Dark leafy green vegetables	61.08	61.80	0.855
Other Vitamin-A rich fruits and vegetables	7.88	8.20	0.887
Other vegetables	**73.40**	**82.79**	**0.003**
Other fruits	10.34	13.93	0.188

according to market visits for each of the above identified food groups per income quintile. The χ^2 p-value is indicating if the changes are significant, where we consider a p-value below 0.1 as significant.

For the food groups "Nuts and seeds" (Figure 3.8) and "Dairy products" (Figure 3.9), we see an increase of consumption with higher income quintiles independent from market access. However, the consumption at the 5th quintile – i.e. the richest 20% of the sample – benefits the most from market access, where nuts and seeds and dairy products are consumed by an additional 20%. Vegetable consumption shows a different picture (Figure 3.10). We see that vegetable consumption is decreasing with higher income groups if markets are not visited regularly. Market access is inverting this trend so that higher income groups also continue to consume vegetables.

3.6.3 Robustness checks

Dietary diversity is influenced by a magnitude of different determinants. This study is focusing on production diversity and market access. In order to control for possible correlation between production diversity and other factors, we include a sufficiently strong instrument variable. Further, we include model specifications different from our main Poisson model, a Poisson IV (GMM) model, an OLS model and a Linear IV (2SLS) model. The results indicate different magnitudes; however, similar direction of effects of the covariates were detected. In a

Figure 3.8: Nuts and seeds consumption per income quintiles

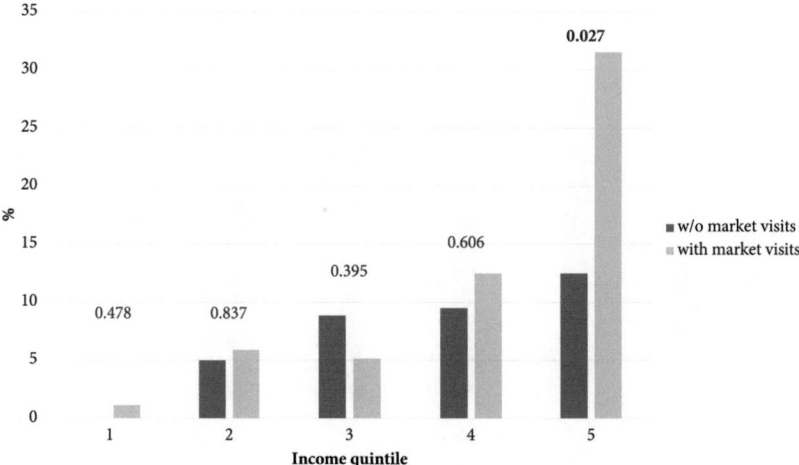

Note: The χ^2 p-values are in red

Figure 3.9: Dairy consumption per income quintiles

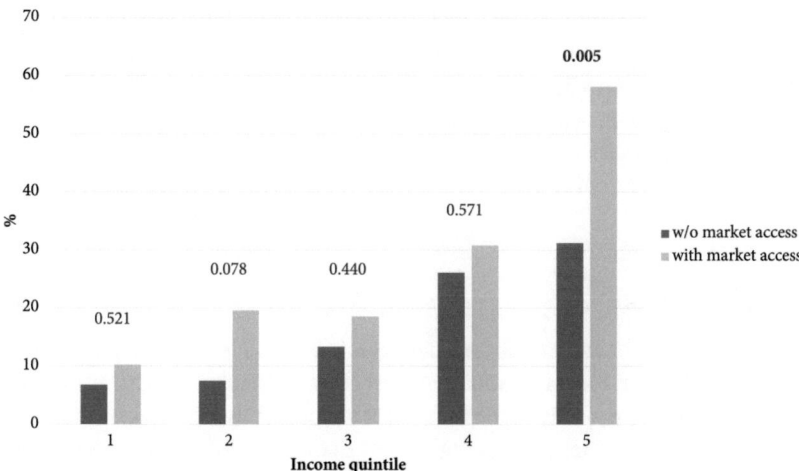

Note: The χ^2 p-values are in red

second set of estimations we use the MDDW, a binary variable, as a dependent variable and as a different measurement for nutrition security. A Probit model uses the same covariates and again displays results with different magnitudes, but

Figure 3.10: Vegetables consumption per income quintiles

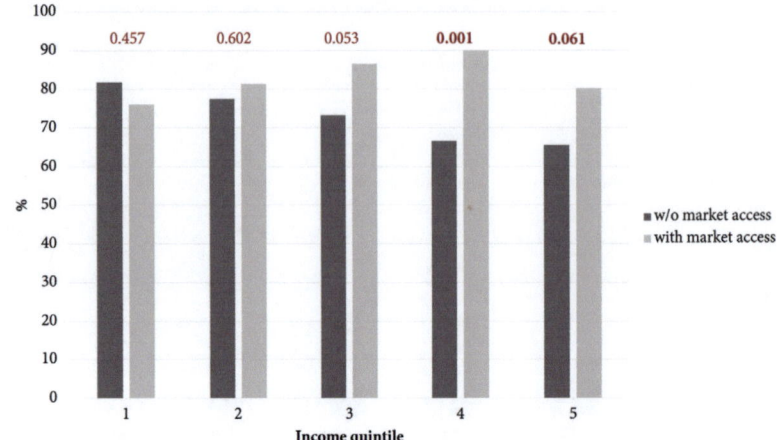

Note: The χ^2 p-values are in red

with the same effect direction. Finally, the interaction is included that once again produces similar results in the effect directions and partially also in effect size. Furthermore, the results are within the expected range given theoretical considerations (see Section 3.3) and based on existing literature (see Section 3.2). Thus, having included a set of model specifications, variations in the dependent variable and other robustness measures such as regional effects, robust standard errors and a tested set of covariates, we interpret the presented results as sufficiently robust.

3.7 Discussion

In the context of the Indian subcontinent, it is indisputable: large parts of the population suffer from food and nutrition insecurity. Particularly at risk are women and children. The majority of the food-insecure people live in rural areas and are smallholder farmers producing a significant share of their food themselves. This study shows that the diversity of production has a positive effect on the diversity of consumption of women in smallholder farmer households. The effects are not highly significant and in the context of similar studies in Sub-Saharan Africa, it seems that the Indian link between production diversity and dietary diversity is likewise not so strong. However, we find a highly significant and much more relevant effect between production diversity and the minimum dietary diversity.

The MDDW depicts the minimum number of food groups needed (namely 5) in order to consume a sufficient number of micronutrients keeping in mind the various limitations of this form of measurement (compare Section 1.5). Hence, production diversity can help to meet this form of nutrition security, but it is probably insufficient to further increase the diversity of diets. The rationale can be found in production economics; the average farmer produces three food groups in one season. Increasing the variety of production would necessitate an increase in profitability. Given that the median production diversity is also three (60% of all farmers produce three food groups), it is probable that this number of food groups is close to optimal production diversity for the farmers of the used sample given the available information and economic environment. Yet, increasing production diversity further significantly improves the probability of adequate micronutrient consumption. Considering the income of farmers, the majority of farmers who produce five or more food groups are from the first two income quintiles and can meet the minimum dietary diversity solely by production. Whereas the majority of farmers who produce three or less food groups are from the last two income quintiles (see Figure B.2 in the Annex on page 191 for a visualization of these correlations). This indicates that richer farmers tend to specialize their production whereas poorer farmers rely more on diversification of their production. It shows that poorer farmers are more likely to rely on their own production for consuming diverse diets. While richer farmers are more likely to rely on markets for consuming diverse diets. This circumstance is further hinted at by the additional results as production diversity is by no means the only channel that affects dietary diversity.

Income is a highly significant and robust factor for dietary diversity, which can be considered in line with the above discussion. The results also indicate that the effects of government programs have a positive effect. We include the number of government schemes that the household actively uses as a count variable without differentiating the programs themselves; yet it significantly increases the diversity of diets. Individual characteristics such as the literacy of women play an important role as well. Household demographics are important in the sense of the household structure; the more male adults that are present, the worse female nutrition becomes. This study cannot make any profound claim on the importance of female empowerment. However, we think that literacy and the number of males can reflect the bargaining power of women within the household, which again represents one additional channel for nutrition improvements.

Market access adds more to the picture. We understand market access not as a mere infrastructural variable as it is often observed e.g. by measuring the distance to markets. Rather market visits are included, which represents the behavior regarding whether markets are visited regularly. We show that market access improves nutrition primarily in regard to food groups that are not necessarily produced on-farm. We cannot infer if market access leads to a certain specialization of on-farm production and thus to a reduction of production diversity, which might explain the results as well. However, we can state that primarily higher income groups benefit from market access. Accordingly, the agricultural household model's separability condition – production choices are independent from consumption choices if households are integrated in markets – does partly hold. Households of various income groups do not differ in mean distance to markets nor in the average behavior of visiting markets, but nevertheless the effects of the markets on nutrition choices differ. Poorer households have a stronger non-separability than wealthier households in the context of our study.

Hence, production diversity increases dietary diversity, but the positive effect of market access can outdo the positive effect of production diversity, although only for higher income groups and only for some food groups. We deduct policy implications as follows:

- Policies and programs for increasing production diversity on farm-level have the potential to be effective for improving nutrition security primarily of lower income groups. Such programs could also target a smaller production than on fields e.g. homestead gardens. However, these might not be the most efficient policies. Policies and programs that have a different goal such as income generation or female empowerment might have a stronger effect on increasing dietary diversity.
- Intensifying the market access can be a viable policy objective for improving the nutrition security if higher income groups are targeted.
- Lower income groups are certainly supported by governmental support programs as these are already existent in India.
- Female empowerment through education policies can have a positive effect on the nutrition status of women.

3.8 Conclusion

This chapter contributes to existing studies on production diversity by answering the research hypotheses, by utilizing a robust methodology and by analyzing the Indian setting. The first hypothesis of this study is verified: On-farm production diversity as measured by the diversity of food groups produced does have a positive effect on dietary diversity. It is not a food group-for-food group relationship.

Instead if production diversity is increased by 1 food group it can be said with high probability that the dietary diversity of women increases up to 18.8%.

The second hypothesis of this study is partly falsified: Market access is not always the channel for this relationship. We see that with market access, the effect size of production diversity decreases, yet remains positive. The effect size of market access is much larger and, in interaction with production diversity, is overall positive. However, market access is primarily benefiting the top income groups, whereas for lower income groups on-farm production diversity remains more relevant.

This study has certain limitations besides the specific ones discussed in-text. Firstly, we see representative results for three distinct rural regions of India; we can by no means extrapolate the results for all of India as the characteristics are too manifold. Secondly, this study uses a cross-sectional survey, which is sufficient to respond to the research questions. Panel data would be superior in controlling for unobserved heterogeneity and for causality claims. Panel data could also shed light on possible seasonality effects and on nutrition variability due to shocks. Thirdly, dietary diversity is the best possible approximation to measuring nutrition security. However, there is still an error margin between dietary diversity scores and actual micronutrient status.

Accordingly, we make the following suggestions for future research. (1) Despite its relevance towards global economics, its population size and its issues with food and nutrition insecurity, India still has a comparatively low level of research results in the research domain of this study. Therefore, we would like to encourage further research activities that include comparative surveys across India. These surveys would be preferably of a longitudinal character and designed as a panel, particularly to control for intra-household effects. (2) Proxy indicators for the micronutrient status such as the DDW or the MDDW can merely indicate approximately if the micronutrient requirements are met. Utilizing biochemical assessments can improve the informative value; hence, the margin of error can potentially be reduced. (3) The role of markets has been studied extensively. However, in regard to food and nutrition security, the depth of analysis is partially lacking. For example, markets can take different shapes in form of size, degree of formality, geographical outreach or accessibility. Most of the literature, including this study, differentiates only marginally if at all. Given the results of differing market effects by income groups and food groups, we see it as relevant to further investigate if informal markets or possible barter trading is of higher importance for the food and nutrition security level of lower income groups in rural areas.

Moreover, the reduction of transaction costs is a decisive factor for efficient markets. Analyzing how this effect particularly applies to food groups with naturally high transaction costs (e.g. fruits and vegetables) can further create insights to the nutrition-enhancing effects of markets. (4) A better understanding of market integration and its links with nutrition choices also opens up the possibility to assess effects of market regulation policies. Subsidies and taxes are policy tools that can steer consumption choices either towards a more frequent consumption of nutritious foods or towards a reduced consumption of unhealthy goods (Franck et al., 2013). Demand effects through price elasticities of certain food groups or food items can be estimated with the available data, however, in a limited realm. A more substantial analysis over time can only be done with longitudinal data and with a better understanding of the local markets. This would enable the estimation of possible substitution effects of consumption or of interlinked price effects such as differing effects for net producers or buyers (Barrett, 1999). Natural experiments also present effective ways to assess food price policies and tax effects. These seem particularly feasible in India considering the diverse and frequently changing policy landscape in the different states with various levels of implementation.

4 Considering Preferences for Food Consumption

Abstract: Research shows that numerous preferences affect economic behavior: attitudes towards risk, time preference and altruism are among these. However, few authors discuss theoretically the link between these preferences and food consumption choices. This study aims to provide this link. Risk preference, time preference and altruism are integrated in the expected utility framework where nutritional consumption is explicitly included. Nutrition is regarded as investment good, providing a health pay-off in the future. The future itself is uncertain, as the model includes the possibility of shocks with unknown effects. An optimal nutrition level is derived and the effects of different levels of preferences discussed. The model is generalized for malnourished populations that primarily gain income from physical labor, hence for the majority of rural populations in developing countries. The theoretical discussion on the effects of preferences on food consumption choices provides a starting point for further disentangling the seemingly irrational behavior of individuals and for explaining as-yet unexplained nutrition behavior.

Keywords: microeconomic behavior, risk preference, altruism, consumption choice, nutrition

4.1 Introduction

The previous chapter showed how production and markets can influence consumption choices. More precisely, we showed which food consumption choices individuals make conditional upon a certain set of production choices, as well as given their access to markets. Indirectly, we postulated that the behavior of the consumers is consistent with the empirical results in that an increase in production diversity will necessarily and causally lead to an increase in dietary diversity. Therefore, on the basis of neoclassical economic theory, we claimed that an increase in production diversity increases the utility of consumers by meeting their preferences for more diverse diets. This is an acceptable axiom if the average effect is estimated. But what if preferences differ among individuals and if postulated utility maximization disregards these preferences?

Normative economic theory postulates that the preferences of individuals determine their consumption behavior. An individual maximizes her utility according to her preferences; thus, rational behavior guides an individual to find the optimal outcome subject to various conditions. Inversely, consumption behavior can reveal the ordinal utility function and the consumers' preferences (Houthakker, 1950; Samuelson, 1938). Utilizing consumers' preferences is at the intersection of positive economics and normative economics. On the one hand, the demand for a choice of goods reveals preferences and is at the same time an indicator for scarcity or surplus. On the other hand, in the best case, policy makers consider revealed consumer preferences for adjusting policy decisions accordingly. However, for a long time, preferences have been considered as exogenous to an individual and as fixed to a status quo; the *homo oeconomicus* as a fully rational agent with distinct preferences is the basis of this axiom. Behavioral economics is changing this picture.

Research in psychology and economics shows that preferences are not only endogenous to the environment, but also an explanatory factor for seemingly irrational behavior of individuals. Preferences are dependent on reference points (Koszegi and Rabin, 2006). The *Endowment Effect* exemplifies this feature; individuals tend to prefer a good that they already own over a good with similar characteristics that they do not own (Kahneman, Knetsch, et al., 1990). Specifically, individuals require a higher price for selling the good in their possession than the amount they would spend to acquire a similar good. *Nudging* as an additional example explains that individuals tend to choose a default option disproportionately more frequently than an alternative option that requires only minimal additional effort (Thaler and Sunstein, 2008). Preferences are also adaptive (von Weizsäcker, 2011, 2015). The consumption behavior itself can influence the preference structure of consumers.

These insights consider preferences foremost as a choice of goods. But preferences are more than just a bias towards a certain set of goods. Deep preferences are at the core of human behavior. Psychology has discussed various mostly subconscious personality traits as influential to a wide range of human behaviors and social actions. The five-factor model identifies core traits as openness, conscientiousness, agreeableness, extraversion, and neuroticism (Digman, 1990; McCrae and Costa, 1987; Rotter, 1966). Similarly, economic theory has adapted certain preference traits in explaining economic behavior. Most notably discussed are time preference, risk preference, altruism, trust and reciprocity (Falk et al., 2018). These deep preferences are used frequently to adjust rational explanations e.g. for future discounting, investment choices, remittances or more conceptually broad structures such as social capital. Yet, little work has been done in the economic literature on linking deep preferences with consumption preferences in order to identify how core preference traits might influence consumption choices.

Even in situations of extreme scarcity, individuals make decisions on what to consume. For example, in food insecure conditions under which the accessibility to food is drastically reduced so that large parts of a population suffer from undernourishment, households tend to decide not to share any food with non-household members (von Braun et al., 1999). This might be considered as irrational behavior because the informal social-support networks including food sharing would on average let all households be better off than non-sharing. There are at least two economic preferences that can explain this behavior. First, present-bias individuals value current consumption more than future consumption. A potential famine increases discount rates close to one, the current survival is the primary need. An informal social-support network only pays off in the future and is therefore irrelevant unless it can be utilized immediately. Secondly, altruism or reciprocity differs depending on the reference group. Sharing of food with the closest relatives is more decisive than sharing with other people.

This narrative exemplifies how deep preferences can affect actual food consumption of individuals and their peers, and consequently, the food and nutrition security situation in general. The exploration of estimating the effects that specific preferences have on nutrition seems relevant. Therefore, this chapter discusses the theoretical considerations of linking deep preferences as identified by economic theory with consumption choices. We use the theoretical framework of expected utility maximization as our basis and expand it with additional preferences of interest: risk preference, time preference and altruism. The aim is to find a hypothesized optimal nutrition level that can predict food consumption choices subject to the preference traits of risk preference and altruism.

4.2 Related Literature and Contribution

The literature review on food and nutrition security suggests that quantifying observable determinants and their impact on nutrition outcomes is a popular research theme. Yet the explanatory power of the models using microdata remains low for these studies. This is not surprising given the nature of the data and modeling approaches; it is, however, surprising that the large share of latent factors that potentially determine nutrition outcomes are not nearly as much studied, particularly not in the setting of developing countries. Thereby revealing the importance of latent factors such as personality traits can create a better understanding for consumption decisions. Vermeir and Verbeke explain these factors as follows: "Everyday consumption practices are still heavily driven by convenience, habit, value for money, personal health concerns, hedonism, and individual responses to social and institutional norms, and, most importantly, they are likely to be resistant to change" (Vermeir and Verbeke, 2006, p. 170).

The economic literature discusses the effect of individual preferences on decision making from two perspectives. On the one hand, preferences are understood as preferences for a certain choice of goods; hence, the goal is to reveal these preferences for policy making purposes. This understanding of revealed preferences tends to regard the preferences as fixed and exogenous. On the other hand, preferences are seen as intrinsic personality traits that lead individuals to make certain economic decisions. As these preferences are influenced and formed through the interaction with and relative to other individuals, these preferences are regarded as endogenous and changeable over time. Some scholars try to bridge the two views and postulate adaptive or induced preferences that are essentially endogenous consumption preferences. The present study builds upon intrinsic personality traits and disentangles the effects of endogenous deep preferences.

Risk

Risk preferences are tendencies to avoid or to engage in actions that might have a future higher return and at the same time a higher default rate (Binswanger, 1981; Dohmen, Falk, Huffman, et al., 2011). Risk tendencies are individual preferences that can explain production choices and investments (Chetty and Szeidl, 2007; Gatzweiler and von Braun, 2016; Liu, 2013), but, as the theory suggests, also consumption behavior (Lusk, Roosen, et al., 2011). Food safety and the willingness to accept possible risks with the consumption of unhealthy foods are mostly associated with a higher risk acceptance (Anderson and Mellor, 2008). Food purchase intentions are influenced by risk perception and trust in food safety information (Lobb et al., 2007). Also, risk tolerance is often positively correlated with risky behaviors such as smoking and drinking (Barsky et al., 1997). However, all studies have in common that risk preference only explains a small fraction of the variation in behaviors.

Regarding nutritious food consumption, Rieger (2015) found for the Cambodian setting that the risk preference of parents does have an impact on the nutrition outcome of their children. Other supporting empirical studies are scarce, primarily due to the lack of proper risk assessments in regard to nutrition (e.g. Fox, 2011; Lusk and Coble, 2005). Studies in rural areas with smallholder farmers focusing on dietary quality are not even existent to the best of our knowledge at the time of writing.

Considering the reliance of poor people on their human capital as the main productive asset, risk preference could influence the dietary quality particularly of such individuals (Bleakley, 2010; Kakietek et al., 2017). This view entails the competing aspect of optimal investment decisions in other capitals and related expected returns, contrary to the view that nutrition is solely a necessary consumption for survival and, as such, any spending on nutrition is not strictly competitive to other investments. The assumption is rooted in the expected utility theory: an individual tries to maximize the net present discounted value of her utility function subject to a limited budget.

Time

Time preference or *time discounting* affect intertemporal choices at which a decision involves tradeoffs between costs and returns at various points in time (Frederick et al., 2002). Time has to be considered, as hidden hunger leads primarily in the long run to negative impacts on health (see Chapter 1 on page 35). Therefore, the second influential aspect proposed is time preference or the preference for discounting. Time preference is commonly used for explaining the relevance of certain economic variables, e.g. saving rates. The intuition is that a present-bias individual has a high discount rate for future periods, which results in low saving rates. The reverse argument holds true for forward looking individuals (compare Deaton, 1992). A few studies have also found an impact of time preference on nutrition outcomes e.g. the BMI (Borghans and Golsteyn, 2006). A systematic review by Barlow et al. (2016) found that 19 studies have shown an impact of time preference on various nutrition variables: among others, on unhealthy diets. For example, it is estimated that an increasing marginal rate of time preference has led to an increase in obesity in the USA (Smith et al., 2005). However, all studies only observed high-income societies and only estimated the impact on unhealthy behavior. They found that time preference is mostly, in interaction with risk preference, a viable predictor for certain nutrition behavior (Tanaka and Nguyen, 2010). Therefore, it is proposed to also include time preference as a variable that should be assessed with this research.

Altruism

Extensive research in behavioral economics found that social preferences are correlated with cooperative behaviors in various aspects of life (Falk, Becker, Dohmen, Huffman, et al., 2016). Altruism is understood as the tendency to unconditionally share part of one's own consumption possibilities with others (Becker, 1981). This behavior has been particularly discussed in a family setting

(Becker and Barro, 1986; Stark, 1995). In larger unrelated groups, social connections have an effect on productivity considering friendships (Bandiera et al., 2005, 2009) or ethnicities (Hjort, 2014). In even larger groups with no direct interaction, the term *warm glow* has been introduced in reference to altruistic behavior (Andreoni, 1995), expressing a general good feeling because of giving if an individual tends to act in an altruistic manner. The motivation to be proactive for creating social interactions and for participating in groups (i.e. in the form of positive reciprocity and altruism) can very well be linked to nutrition behavior. This holds true particularly for the nutrition behavior of children who are dependent on their parents that have a certain social preference and related behavioral characteristics. There is an intergenerational survival value in situations of food scarcity: parents share with their children and in return their children are more capable and healthy for participating in income generation and eventually for reproduction (Bergstrom and Stark, 1993). Conceptually, altruism is closely linked with reciprocity. *Positive reciprocity* is the willingness to share while having expectations of a positive return. Conversely, *negative reciprocity* is the willingness to retaliate. Particularly, in non-kin groups, reciprocity might be the actual motivating factor for certain actions due to self-interest (Charness and Rabin, 2002; Forsythe et al., 1994). Nonetheless, the outcome of altruism and reciprocity is the same considering one point in time and from an individual perspective. Accordingly, we will analyze how altruism influences nutrition as a third preference.

Endogeneity of preferences

The endogenous character of preferences needs to be emphasized. Preferences are changing over time not only in terms of consumption choices but also in terms of personality traits. Macroeconomic shocks can affect risk taking (Malmendier and Nagel, 2011). Also, aging and health shocks can affect the stability of preferences over one's life span (Dohmen, Falk, Golsteyn, et al., 2017; Hideki, 2013). Furthermore, there are clear differences of preferences between genders (Croson and Gneezy, 2009; Falk and Hermle, 2018). Generally, endogeneity forms a challenge in behavioral economics for two reasons: from an empirical perspective, since the true effect of a preference on a decision is statistically difficult to estimate; and in the form of policy implications, as each recommendation that follows from a study has the potential to change the preference profile of a given population, hence, the effects of policy implications can potentially undermine their intention. We acknowledge the challenge presented for this research and interpret the results accordingly. However, for the theoretical model we consider that in a given economic setting, preferences are taken as given (Stigler and Becker, 1977).

Based on the discussion of preferences in economics, this study contributes by discussing theoretically the linkages between preferences and individual nutrition considering the impact on food and nutrition security. The research objective of this chapter is *to analyze the effects that risk preference, time preference and altruism have on nutrition choices under conditions of future uncertainty.*

4.3 Theoretical Model

Intuition of the approach

To understand the intuition of our approach, let us consider a classical life-cycle model where individuals make decisions between consumption and savings (e.g. Deaton, 1992, Chapter 6). Income is the product of wage and working hours. Given one individual or household, the wage level is exogenous and only the amount of working hours are endogenous, being dependent on the capability and decision of an individual. An individual would thus increase her working time if she either wants to increase current consumption or her savings. Savings can have a precautionary motive to mitigate future uncertainty in an intertemporal allocation problem (Deaton, 1992, p. 177). Preferences affect this decision. Risk averse individuals would increase current working time in order to increase the savings. Present-biased individuals would increase the consumption-to-savings ratio. Altruistic individuals would have a lower current or future consumption because of donations.

Keeping these linkages in mind, we can amend the basic model by including consumption of nutrition. Nutrition can be understood as an investment decision because a better nourishment today likely results in better health in the future, which in turn increases the available working hours or productivity in the future. However, risk averse individuals would increase savings, which limits the resources available for nutrition. Thus, risk averse individuals would likely allocate a suboptimal budget for nutrition unless perfect knowledge on the returns of nutrition exists. Accordingly, a theoretical model that takes into account preferences and their effect on future outcomes should find the optimal level of nutrition considering savings.

Theoretical frame

The theoretical framework for this study is rooted in the demand for health model as introduced by Grossmann (1972). We diverge from Grossmann particularly in regard to the concept of health capital, but would like to present this essential work nonetheless.

Grossmann states that each individual should be regarded as a producer of her health. As such health can be regarded as a form of human capital. The health capital stock produces an output of healthy time, which is used to earn income. Health capital depreciates over time but can be increased by investment e.g. by medical care or nutrition. According to Grossmann, health capital differs from other forms of capital in that the health capital stock determines the available time for utility whereas other forms of human capital determine the productivity levels. Therefore, human capital is the function determining time to earn income: the healthier one is, the longer the income accumulation. We recognize that human capital that entails health capital will actually also affect productivity level and not only the available time for productive activities, but for now an individual acts to maximize the intertemporal utility of health and other goods (Grossman, 1972, p. 225):

$$U = U(\phi_0 H_0, ..., \phi_n H_n, Z_0, ..., Z_n),$$

where H_0 is the initial (inherited) stock of health, H_i is the stock of health in the *i*th time period, ϕ_i is the service flow per unit stock, $\phi_i H_i$ is total consumption of health services and Z_i is the total consumption of another commodity in the *i*th period. n depicts the length of life, which is endogenous. n depends on the quantities of H_i that maximize utility subject to certain production and resource constraints.

Grossman introduced his model particularly as an extension to the utility maximization of Becker (1967) and Ben-Porath (1967) to determine an optimal investment in human capital at any time. These models are assuming a perfect knowledge of the future, an assumption that was challenged by Levhari and Weiss (1974) in the context of human capital. Levhari and Weiss introduced the expected utility maximization that includes risk on future outcomes:

$$\max_{c_0, \lambda} V = E\{u(c_0, c_1)\},$$

where λ is the investment share in human capital, c_0 is consumption in *period* 0 and c_1 is consumption in *period* 1. An individual maximizes the expected value of lifetime utility from consumption by considering an unknown future, i.e. uncertainty about education or exogenous individual abilities as well as the uncertain future earning capacity of human capital due to imperfect knowledge of future labor demand and supply. This utility function is subject to the wealth constraint:

$$c_1 = (A + (1-\lambda)y_0 - c_0)(1+r) + y_1$$

with

$$y_1 = f(\lambda, \mu),$$

where A is initial wealth, y_0 are earnings in *period* 0, and r is the market rate of return. Notably, y_1 is future earnings and a function of the investment in human capital λ and of the future uncertain state of the world μ, which is a random variable with known distribution.

4.3.1 Preference model for nutrition

We would like to diverge from the literature and consider particularly the nutrition of an individual as an intertemporal decision making process. We also consider that nutrition should be regarded as an investment in human capital that influences future income. Nutritional intake in the present leads to a certain health outcome in the future, which in turn is the basis for income generation in the future. However, we diverge from Grossmann (1972) in our model in that the outcome of nutrition does not change the number of time periods available for utility. We consider only two periods, *period* 0 as present and *period* 1 as the future.

Individuals earn income and invest their income either in nutrition or any other consumption items; residual income is considered as savings for the future period. Accordingly, consumption decisions in the present are a direct trade off between savings, nutrition and other expenditures. Future earnings depend on the investment in nutrition during the present and future wage rates. However, the future is uncertain. Current nutrition can only generate an uncertain health outcome (refer to the UNICEF framework with linkages from nutrition intake to nutrition outcome on page 31) and - possibly more challenging - the future might bear shocks that affect individual livelihoods[1]. The model has to include an assumption for the nutritional outcome, which considers diminishing marginal returns of nutrition. The assumption is formalized at a later step below.

Literature on behavioral economics considers various preferences as influential to individual decision making. At this point, we include two individual

---

1   Shocks can be positive or negative. The frequency and magnitude of shocks is unknown, but is assumed to be normally distributed.

preferences as decisive: risk preference, which can be a guidance on how individuals judge the magnitude and frequency of future shocks; and time preference or personal discount rates that value the expected utility of the future differently. Hence, we consider a Neumann-Morgenstern expected utility function[2]:

$$V_{c_0,c_1} = u(c_0) + \int_{-\infty}^{\infty} \delta u(c_1)\varphi(x)dx, \qquad (4.1)$$

where utility in the first period is dependent on consumption c_0, utility in the future period is dependent on consumption c_1, the discount factor δ (with $0 < \delta < 1$) and on a probability density function $\varphi(x)dx$ of the uncertain future with shock x. The individual maximizes the expected utility along nutrition and savings in the present:

$$\max_{n_0,s_0} V_{c_0,c_1} = E\{u(c_0,c_1)\},$$

where the utility function is strictly monotone and concave. The maximization is subject to the budget constraint:

$$c_1 = y_1 + (1+r)s_0, \qquad (4.2)$$

where the consumption c_1 of *period* 1 is a function of the earnings y_1 in *period* 1 and the savings s_0 of *period* 0 multiplied by the market rate of return r. Savings are earnings of *period* 0 subtracted by the costs for nutrition n multiplied by price p and other consumption c_0 of *period* 0:

$$s_0 = y_0 - n_0 p_0 - c_0$$

from which we also get the budget constraint for *period* 0:

$$c_0 = y_0 - n_0 p_0 - s_0 \qquad (4.3)$$

Further, as future earnings y_1 in *period* 1 are directly affected by nutrition in *period* 0, we can understand earnings y_1 as wages w multiplied by the uncertain outcome of nutrition, represented as a function $f(n)$ and multiplied by the expected mean of the stochastic shock μ. As the reader can see, we explicitly diverge from Grossmann (1972) as well as from Levhari and Weiss (1974) in that we recognize the outcome of nutrition as a determinant of the level of production (contrary to Grossmann) and that the future income is not only dependent on the investment in human capital, but also on an exogenously given wage

---

[2] We follow Rieger (2015) in setting up the model, but we explicitly diverge from Rieger in finding optimal points.

rate (contrary to Levhari and Weiss) and the uncertain future. Accordingly, we propose:

$$y_1 = w_1 \mu f(n_0) \tag{4.4}$$

Substituting equation (4.4) in the preliminary budget constraint (4.2), we get the budget constraint for *period* 1:

$$c_1 = w_1 \mu f(n_0) + (1+r)(y_0 - n_0 p_0 - c_0) \tag{4.5}$$

Substituting (4.5) in (4.1), we get:

$$V = u(y_0 - n_0 p_0 - s_0) + \int_{-\infty}^{\infty} \delta u \big[w_1 \mu f(n_0) + (1+r) s_0 \big] \varphi(x) dx \tag{4.6}$$

For empirical applications, we want to specify the expected utility model. We assume the utility as a negative exponential function and include the measure of risk preference A for reflecting the individual expectation of a risk, which the literature specifies as[3]:

$$V_{c_0,c_1} = -e^{\{-A c_0\}} - e^{\{-A c_1\}}$$

Considering the above specification for equation (4.6), we obtain:

$$V = -e^{\{-A(y_0 - n_0 p_0 - s_0)\}}$$

$$- \int_{-\infty}^{\infty} \delta e^{\{-A[w_1 \mu f(n_0) + (1+r) s_0]\}} \varphi(x) dx$$

We rewrite the integral in a mean-variance form with the assumption that stochastic shocks are normally distributed with an unknown mean and an unknown

---

[3] Note that the Arrow-Pratt measure of absolute risk aversion is used for the interpretation of the model (Arrow, 1965; Pratt, 1964). An individual is risk averse if $A > 0$, risk neutral if $A = 0$ and risk seeking if $A < 0$, where A is the measure of risk aversion. We choose this particular specification of an exponential utility primarily because it implies constant absolute risk aversion, i.e. the level of risk aversion does not change with respect to wealth or income, formally: $A(c) = -\frac{u''(c)}{u'(c)}$. This is useful for finding an optimal nutrition level that is independent from other consumption and savings, but dependent on levels of individual preferences. Also note that the interpretation of the expected values of the exponential utility are interpreted ordinally instead of cardinally.

variance (see Appendix on page 193 for the solution of the integral), so that we get the final form for which individuals maximize their utility[4]:

$$\max_{n,s} V = -exp\{-A(y-np-s)\}$$
$$-\delta exp\left\{-A\left(f(n)\mu w + (1+r)s - \frac{1}{2}Af(n)^2\sigma^2 w^2\right)\right\}, \quad (4.7)$$

where A is the absolute measure of risk preference, s are the savings in *period 0*, y is the income in *period 0*, n is the nutrition in *period 0*, accordingly $f(n)$ is a function of the nutrition, i.e. the nutritional outcome in *period 1*, δ is the personal discount rate, r is the market rate of return, w is the wage rate in *period 1* and the normally distributed shocks are denoted with the mean μ and the variance σ^2.

Our aim is to solve the utility function with an optimal nutrition level, in which the relationship to preferences can be identified. From an economic point of view, risk preference would primarily affect saving rates as a means of mitigating possible future shocks. To disentangle this effect in the model, the optimal nutrition rate needs to be solved as independent from savings. Therefore we need to find the maximum of the utility function over nutrition and savings, for which we derive the two First Order Conditions (FOCs) and by substituting the optimal savings rate in the equation of the optimal nutrition rate with respect to nutrition n and savings s (see Appendix on page 195 for the differentiation of the FOCs)[5]:

$$\frac{dV}{dn} = -Ap\,exp\{-A(y-np-s)\}$$
$$+ (A\mu w f'(n) - A^2\sigma^2 w^2 f(n)f'(n))$$
$$* \delta exp\left\{text-A\left(f(n)\mu w + (1+r)s - \frac{1}{2}Af(n)^2\sigma^2 w^2\right)\right\}$$

---

[4] In the following we rewrite the exponential function with *exp* and we delete the time subscripts for better readability.
[5] For didactic reasons we use d for the partial derivatives instead of δ since the preference model includes δ as the discount rate. MATLAB R2018A was used for all further differentiations.

and

$$\frac{dV}{ds} = -A\exp\{-A(y-np-s)\}$$

$$+A(1+r)$$

$$*\delta\exp\left\{text-A\left(f(n)\mu w+(1+r)s-\frac{1}{2}Af(n)^2\sigma^2 w^2\right)\right\}$$

By finding the maximum at $\frac{dV}{ds} = 0$, we can solve for s, which will give the optimal savings rate s^* (see Appendix on page 199):

$$s^* = \frac{A(y-np)+\frac{A^2 f(n)^2 w^2 \sigma^2}{2}-Af(n)\mu w}{2A+Ar}$$

We substitute s^* in the maximum at $\frac{dV}{dn} = 0$, from which one can easily derive the preliminary step for the final solution, the implicit solution of the optimal (maximum) nutrition level n^* that is independent of savings s^*:

$$\frac{dV}{dn^*} = -Ape^{\left\{A\left(np-y+\frac{A(y-np)+\frac{A^2 f(n)^2 w^2 \sigma^2}{2}-A\mu f(n)w}{2A+Ar}\right)\right\}}$$

$$+(Af'(n)\mu w - A^2 f'(n)f(n)w^2\sigma^2)$$

$$*\delta e^{\left\{\frac{A^2 f(n)^2 w^2 \sigma^2}{2}-\frac{A\left(A(y-np)+\frac{A^2 f(n)^2 w^2 \sigma^2}{2}-A\mu f(n)w\right)}{2A+Ar}\right\}}$$

$$*e^{\left\{-Af(n)\mu w-\frac{Ar\left(A(y-np)+\frac{A^2 f(n)^2 w^2 \sigma^2}{2}-A\mu f(n)w\right)}{2A+Ar}\right\}}$$

$$= 0$$

The solution is implicit due to the unknown function of the nutritional outcome $f(n)$ and prevents an explanation that is meaningful at this point. Accordingly, we are aiming for an explicit solution. We can approximate $f(n)$ logically by stating that this function is a power function with diminishing returns of n. For the purpose of this study, we assume a standard power function with an exponent of $\frac{1}{2}$, such that:

$$f(n) \mapsto n^{\frac{1}{2}}$$

By substituting $f(n)$ in $\frac{dV}{dn^*} = 0$, we get an optimal nutrition level that can be solved for n:

$$\frac{dV}{dn^*} = -Ape^{\left\{A\left(np-y+\frac{A(y-np)+\frac{A^2 nw^2\sigma^2}{2}-A\mu\sqrt{n}w}{2A+Ar}\right)\right\}}$$

$$-\left(\frac{A^2 w^2 \sigma^2}{2} - \frac{A\mu w}{2\sqrt{n}}\right)$$

$$* \delta e^{\left\{\frac{A^2 nw^2\sigma^2}{2} - \frac{Ar\left(A(y-np)+\frac{A^2 nw^2\sigma^2}{2}-A\mu\sqrt{n}w\right)}{2A+Ar}\right\}}$$

$$* e^{\left\{-\frac{A\left(A(y-np)+\frac{A^2 nw^2\sigma^2}{2}-A\mu\sqrt{n}w\right)}{2A+Ar}-A\mu\sqrt{n}w\right\}}$$

$$= 0$$

Solving for n yields the optimal nutrition level in *period* 0, indicated as n^*:

$$n^* = \frac{\delta^2 \mu^2 w^2}{(A\delta w^2 \sigma^2 + 2p)^2} \quad (4.8)$$

where δ is the personal discount rate, A is the absolute measure of risk preference, μ is the mean and σ^2 is the variance of normally distributed shocks, w is the wage rate, and p is the price of nutrition. Hence, equation (4.8) indicates the level of nutrition in the utility function (4.7) that yields the maximum utility over all time periods and that is independent of savings[6]. This is the final solution for this exercise. In the following, we will discuss the model on the basis of equation 4.8 and clarify the implications of the linkages.

---

6   The optimal nutrition level is regarded as a balanced diet that includes all nutrients as recommended by WHO. As a proxy, we use dietary diversity scores (compare section 1.5). Further, in the theoretical model only the individual's perspective is regarded. Nevertheless, we can make assumptions of other individuals in the same household since food consumption is in the specific setting of the research area a communal activity. As such, we regard primarily altruism as an important preference, which is included in an extended model below.

4.3.2 Discussion of the model

First, the model is specified such that nutrition could theoretically become zero. In case of no nutrition, the income of the future period will be zero, intuitively because an individual risks dying without nutrition. Realistically, an individual can survive for a certain period of time without nutrition; likewise, an individual can consume food items that the individual received for free. These options are not considered in the model, and given the utility maximization with a certain income in *period* 0, the optimal nutrition level will always be positive (unless all specified variables in equation (4.8) will be zero, which is unlikely).

Market factors

The relevant market factors are the wage rate w as well as the price of nutrition p; both are exogenous to an individual. Keeping everything else equal, a positive change of the wage rate will induce a higher investment in nutrition. A higher investment in nutrition will lead to a higher nutrition outcome in the future, which in turn increases the ability (or productivity) to generate income and respectively increases the available budget in the future. Similarly, an increasing price of nutrition has the opposite effect. A higher current price naturally reduces the food resources that an individual can consume given the budget constraint. Accordingly, with increasing prices the optimal nutrition level in the present decreases. Although these are relevant effects, for the purpose of this study we will not test these two findings empirically.

Shocks

Unknown shocks are denoted by their distribution with mean μ and the variance σ^2, the full specification is relevant for the optimal nutrition level, although with varying effects. The greater or smaller the mean is, the higher the nutrition in *period* 0 will be. Whereas the greater the variance, the smaller the nutrition in *period* 0 will be. Figure 4.1 on the next page clarifies this intuition. Note that a mean of 0 would cancel out any optimal nutrition level, this is due to the model specification that explicitly links expectations of shocks with future income. However, an expectation of shocks with a mean of 0 would partly include the expectation of no shocks, which is unrealistic in risky rural environments for individuals that are predominantly active in agriculture-related activities.

Figure 4.1 on the following page shows four different normal distributions of hypothetical shocks. Considering the red curve with a mean of -3 and comparing it with the other curves with mean of -1. The red curve indicates that in 99% of

Figure 4.1: Probability density functions of shocks with varying mean μ and variance σ^2

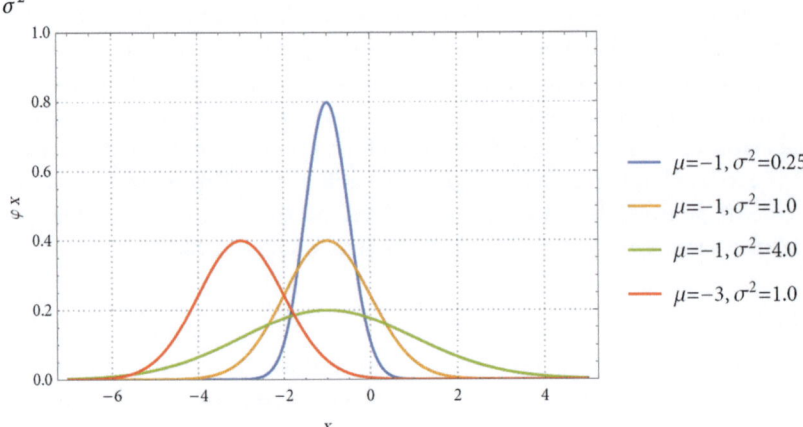

Note: Simulated with Mathematica 11.3.0

the times, the future shock x will be negative; thus, a shock will mostly always negatively affect the available budget of an individual. Accordingly, keeping everything else constant an individual considers an uncertain future as possibly devastating and, thus, increases her consumption in the current period. If we would not keep everything else constant, this consumption effect will be balanced with an increase in savings depending on individual risk and time preferences. In case the mean would be positive, an individual would in fact expect positive effects on the future available budget, and, hence, would also increase the nutrition in *period* 0.

Comparing the blue, yellow and green curve, which have an equal mean of -1 but increasing variances, we can see that the probability of a certain shock to occur decreases with increasing variance, but also the sum of all probabilities for expected positive shocks (respectively more positive shocks) increases. Accordingly, the expectation of an individual would be a future with on average more positive events, and the optimal nutrition level in the present would decrease. However, given equation (4.8), the effect size of a change in the variance is much smaller relative to the change in the mean.

Time preference

Time preference δ has a positive effect on the optimal nutrition level. The intuition is clear: present-biased individuals who discount the future will increase their consumption and nutrition in the present. The discount rate δ is measured as a positive number between 0 and 1. A δ of 0 would entail an equal valuation of present and future, which is non-existent from a consumption point of view. In general, discount rates tend to be at the high end, i.e. $0.7 < \delta \leq 1$. Marginalized and poor households tend to have a δ leaning towards 1, indicating an even stronger preference for current consumption.

Risk preference

Risk preference denoted as the measure for the absolute risk aversion A stands in the denominator, and although the denominator is squared, the direction of risk preference matters, keeping in mind that the Arrow-Pratt measure of risk preference is defined as risk averse if $0 > A \leq 1$, risk neutral if $A = 0$ and risk seeking if $-1 \geq A < 0$. Analytically, we can expect that risk loving individuals will have a higher nutrition level than risk neutral or risk averse individuals. This is a negative relationship, which is non-linear and convex, though the form of the relationship partly also depends on the value of the other variables in equation (4.8). Intuitively it can be explained as risk loving individuals in comparison to others are willing to consume/invest more in the current period with the prospect of higher returns (or losses) in the future. Hence, investment in nutrition in the current period also increases, as does the optimal nutrition rate for risk loving individuals. Figure 4.2 on the next page visualizes the sensitivity of the relationship holding all other variables constant. Three curves are displayed to indicate the effect of risk preference conditional on varying discount rates. Two observations need to be noted: *first*, the higher the discount rate (the more present-biased and individual is, the bigger is the effect of risk avoidance on nutrition. The intuition is that a stronger present-bias incentivizes an individual to consume more in the current period than to save for the future. *Second*, the correlation between risk preference and nutrition is nonlinear; hence, more risk averse individuals have a relatively stronger effect on their nutrition.

We can deduct a number of hypotheses from the model. However, for the purpose of this study and due to the available data, we would like to discuss the following hypothesis, which we will empirically investigate in the following chapter:

> *Hypothesis 1*
> Risk preference measured as an individual level of risk aversion positively affects the current nutrition level of individuals.

Figure 4.2: Effect of varying risk levels A on the optimal nutrition level n^*

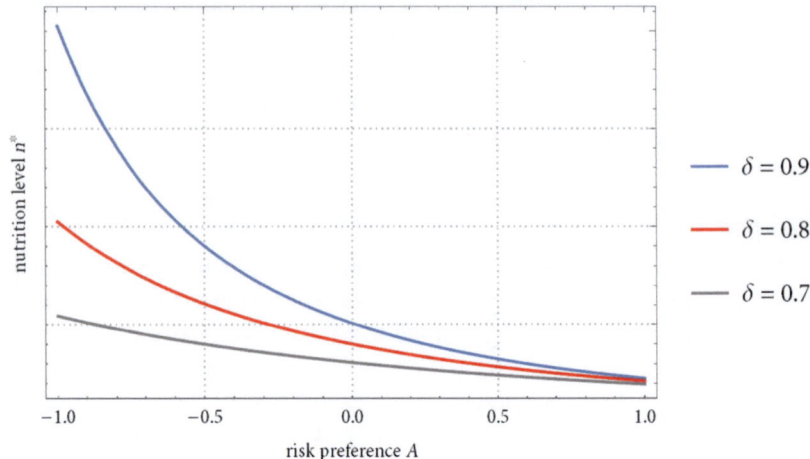

Note: Simulated with Mathematica 11.3.0

4.3.3 Model extension for altruism

Individuals usually live in communities or households, where it is common to share various items, among these being food items. A preference that has been studied quite frequently in behavioral economics is altruism or respectively the preference for sharing one's own items with others solely for the purpose of sharing. Sometimes the purpose of sharing and its sensation is called a *warm glow*[7]. The above model is quite suitable to also include altruism as an additional preference, and the study context of this research shows that altruism is a relevant aspect of social behavior particularly for food consumption[8].

---

[7] Disentangling the effect of true altruism, i.e. donations without expectations of returns other than a warm glow, require a specific experimental design (see Fehr and Schmidt, 2006). In most cases altruism cannot be differentiated from positive reciprocity, which is giving in expectation of returns from the receiver of the giving. The proposed model will not differentiate between altruism and positive reciprocity since the effect of both behaviors should be strongly positively correlated. The robustness checks of the next chapter will consider a substitution of both preferences and indicate the alignment of the effect's direction.

[8] The argument is specific to the cultural setting in South Asia that this research focuses on. Generally speaking during meals, the household head eats first before anyone else in the household eats. The spouse eats last after everyone else.

Theoretical Model

We follow Becker and Barro as well as partly Stark (Becker, 1981; Becker and Barro, 1986; Stark, 1995) for including altruism in the proposed utility function (4.7), whose approach is still today a commonly used benchmark model. We take as given that an individual is effectively altruistic towards another family member. *Altruistic* means that the general utility of an individual depends positively on the well-being of another individual within the same household. *Effectively* is here understood as meaning that an individual's altruism actually changes the utility of the other family member. In line with Becker and Barro (1986, p. 70), the utility V_i of an individual i in a one period model would then be given by:

$$V_i = v_i(c_0) + \alpha_i(V_j),$$

where v_i is the standard utility function (with $v_i' > 0, v_i'' < 0$) that is dependent on i's own consumption c_0 with $c_0 > 0$. V_j is the utility of another household member j. The term α_i measures the degree of altruism towards another household member j and converts the utility of j into that of i. Along Stark (1995, p. 16), we assume that $0 < \alpha_i < 1$, meaning the individual i is neither masochistic nor envious.

Specifically, altruistic in relation to nutrition means that an individual's own consumption will be *reduced* the more altruistic the individual is because part of the available food items would be shared with another individual. For a naive model, two assumptions need to be upheld: (1) utility generated from nutrition is equal for two different individuals in the household, and (2) the food items available to one individual is independent from the food items available to another individual. Considering for reasons of simplicity only *period* 0 and considering that the level of altruism of a household member, say, the male household head that is the husband, may positively affect the nutrition of another family member j, say, his spouse, the model can be formally described as:

$$V_j = v_j(c_0) + v_j(\alpha_i n_i)$$

where α_i is the altruism level of another household member (e.g. husband) other than the specified member j (e.g. spouse), n_i is the available nutrition of i, and $v_j(\alpha_i n_i)$ is the partial utility that j receives from the shared nutrition.

Considering the utility model (4.7), we would include $\alpha_i n_i$ as the factual change of the nutrition n_j of an individual j that maximizes her utility V. We assume that the increase in nutrition of an individual j is not explicitly the partial consumption of another individual's nutrition n_i, but instead an increase of j's nutrition, which is $(1 + \alpha_i)n_j$. We also assume that the additional nutrition received through altruistic sharing is not expected ex ante (i.e. at the time of the decision on expected events); hence, the additional nutrition is not accounted for in *period* 0, but it will contribute to the function of nutrition in *period* 1:

$$V_j = -exp\{-A(y - n_j p - s)\} \qquad (4.9)$$

$$-\delta exp\left\{-A\left(f[(1+\alpha_i)n_j]\mu w + (1+r)s - \frac{1}{2}Af[(1+\alpha_i)n_j]^2 \sigma^2 w^2\right)\right\}$$

At this point we will not further differentiate for the optimal nutrition level in respect to a maximum utility, but instead recognize that the consumed nutrition of an individual j increases proportionally to the level of altruism of the relevant altruistic individual i. Hence, for a given utility per individual, the extended model (4.9) offers the additional falsifiable hypothesis:

> *Hypothesis 2*
> Altruism of another household member j towards an individual i positively affects total utility of i's nutrition due to the increased amount of food items n_i.

4.3.4 Limitations of the model

The proposed model identifies a basic relationship between preferences and investment in nutrition by considering economic variables commonly used for utility maximization problems. In the following, a few considerations regarding the contribution of the above specified model are discussed, i.e. in regard to the inclusion of nutrition and preferences. With the limitations in mind, we will then proceed with empirically testing the hypotheses.

- The model includes preferences as exogenous to the individual; however, these might be endogenous and related to any included variable. The model tries to dissolve this link by calculating the optimal nutrition that is independent from savings. Nevertheless, particularly risky environments (i.e. expected shocks with strongly negative means and small variances) could potentially increase the risk aversion of individuals.

    A very complex model could simulate preferences as functions of the other included variables, although many assumptions about the form of the relationships would limit the precision and predictability power of the model. Instead, analytics can test for possible linkages (options would include sensitivity and robustness checks or simultaneous equations models).
- From the point of view of an individual in the present, the future wage rate is unknown. An adaptation of the behavior in the present based on an unknown future variable is not possible. However, an individual has expectations about

the future wage rate based on previous experience or other information.
In general, an expected wage rate might be sufficiently close in value to the future real wage rate, although the model could be improved by including uncertainty not only about shocks, but also about wages.
- The model explicitly states that future income is generated dependent on the nutritional outcome. This holds particularly true for individuals, whose productivity mainly depends on physical labor. The fewer activities are physical activities, the weaker this relationship might get.
This model is most suitable in the context of physical labor, which requires related data for the analysis.
- The model assumes that more spending on nutrition leads necessarily to a better nutrition outcome in the future. This assumption can be challenged with three points.
First, spending on food items does not differentiate which food items are consumed. As explained in Chapter 1.4 on page 35, a balanced human nutrition requires a certain set of nutrients, which is specific to the individual. It cannot be assumed that the individual knows which set of nutrients are required, nor if the spending is optimally allocated to this set of nutrients. Hence, the spending on nutrition will likely be inefficient. Furthermore, particularly in underregulated societies, food safety issues that are not apparent (e.g. contamination of basic food items with heavy metals or aflatoxins) are existent. Even if the spending on nutrition would be optimal, possible food safety issues could even harm the health outcome instead of increasing the health outcome in the future.
Second, among others the UNICEF framework for nutrition clarifies the manifold linkages between nutrition intake and health outcome (see Figure 1.2 on page 31). A direct relationship between nutrition intake and health outcome can be postulated, but is necessarily a simplification.
Third, the model misses explicitly integrating knowledge on general nutrition; likewise, utilization of food is missing. The FAO Food Security Framework lists utilization of food as a core feature to secured food (see Chapter 1.3 on page 29).
To test this limitation, an analysis of the actual relationship between nutrition intake and nutrition outcome could reveal the nature of the function.
- The proposed model postulates an optimal nutrition level dependent on certain variables and a strictly increasing relationship for most variables. The model leaves out of consideration possible overnutrition effects such as obesity or diet-induced, non-communicable diseases.

A future adjustment of the model might therefore consider a more complex function of nutrition that itself has a vertex point.
- The model specification allows a zero-investment into nutrition, which potentially could lead to death of the individual in the second period, which clearly reduces the expected utility to zero.

A minimum level of necessary spending on nutrition could be included in the model that sets a certain level of spending, which is independent from any preference. Any spending above this minimum level would then be subject to the discussed variables.

4.4 Conclusion

Personality traits form human behavior. Preferences influence economic decisions. Hence, consumption choices not only reveal the preference for a certain choice of goods, but also consumption choices are much more influenced by a certain preference profile. This chapter discussed the links between economic preferences and food consumption choices. The expected utility maximization framework was amended with risk preference, time preference and altruism. An optimal nutrition level was deduced that indicated a positive relation between risk preference and nutrition, and a positive relation between time discounting and nutrition. The model was amended and indicated a positive relation between the altruism level of family members and nutrition.

The proposed model presents a first attempt to link findings from behavioral economics with intertemporal consumption choices and the effects on nutrition intake. The model is designed with a focus on malnourished individuals that primarily gain income from physical labor. This limits the scope for predictions, but at the same time covers a large part of the population in rural areas of developing countries. The model has further limitations that can be addressed in future research, which are of a technical nature (e.g. minimum levels of spending); other limitations are in regard to its scope (e.g. malnutrition and effects on cognitive labor).

In general, theoretical models can provide the basis for policy implications. Normative economic theory uses preference revelations to design "good" policies that consider consumer preferences. Using the presented model to deduct policy implications is, however, premature. Firstly, policy implications that are deduced from preferences potentially entail preference changing effects. A simulated policy effect cannot consider the true extent of preference shifts, partly because preferences have manifold causes and policies are only one. Secondly, the presented model has yet to be tested. The chapter presents a theoretical exploration, which draws its conclusions on the basis of existing theory and previous findings. This is no guarantee for its accuracy.

Ultimately, the predictions of the model are only as good as the extent to which these can be verified or respectively falsified in an empirical setting. The next step is to apply the model in the field and to test its hypotheses. Considering the suitability of the model, we apply the model in rural India. The Indian setting and the sampling frame provides a population whose nutrition is below Indian average and whose livelihoods are predominantly relying on agriculture and manual work (compare Section 1.6.1 on page 42). The next chapter gives further empirical insights.

5 Effects of Preferences on Food Consumption

Abstract: The economic literature presents risk preference and altruism as core factors that affect individual decision making. Since nutrition affects the health and productivity of individuals, food consumption choices can be considered to be individual investment decisions that enable future income generation. We utilize this link to assess and quantify the effects that individual risk taking behavior and altruism of peers have on food and nutrition security. A household-level survey has been conducted that integrates socioeconomic and nutritional data with the elicitation of preferences through hypothetical games. The sample includes three regions of India and 1177 households from 111 villages. By using multiple regression analysis, a positive association of risk preference on dietary intake is estimated. An increase of 10 percentage points in risk taking increases the dietary diversity score by 0.9% to 1.4%. The effects are confirmed for other indicators of food and nutrition security and are found to be statistically robust. Altruistic behavior of the household head improves the nutrition by 1.1% to 3.0%. This study contributes to the existing literature by analyzing the effects of core character traits of behavioral economics on individual nutrition. We show that generally unobserved preferences can contribute to understanding heterogeneous malnutrition rates in nutrition insecure and risky environments.

Keywords: microeconomic behavior, risk preference, altruism, dietary diversity, food and nutrition security, India

5.1 Introduction

The decision on what to eat and how much to eat is dependent on accessible resources, on personal preferences and on societal choices. This has been widely discussed by various authors (prominently Sen, 1981). Decisions on food consumption as an optimization problem that is influenced by various preferences is not as widely discussed. While a basic set of macro- and micronutrients needs to be met in order to be fully nourished, the combination of these nutrients with the specific food items is generally a personal decision that depends on socioeconomic determinants as well as individual choices. The previous chapter discussed the theoretical framework on which nutrition choices can be modeled considering nutrition as an investment decision. The model gives predictions on food consumption dependent on a set of economic preferences.

This chapter presents the empirical assessment of the model's hypotheses. The effects of risk preference and altruism on nutrition security is quantified and discussed in relation to other influential variables. We show that under uncertainty, individual nutrition decisions differ depending on varying economic preferences.

Classical economic literature proposes to reveal preferences by analyzing economic decisions and, specifically, to uncover individual preferences by measuring the willingness to pay for certain goods or foods (Houthakker, 1950; Samuelson, 1938). These preferences are certainly alterable by the current individual situation or by profane mechanisms such as marketing. The literature in behavioral economics and psychology presents a set of preferences that are intrinsic to individuals and that are less alterable; yet these preferences can have an equal effect on consumption choices (Digman, 1990; McCrae and Costa, 1987). These economic preferences are also referred to as deep preferences. The previous theoretical chapter focuses on time preference, risk preference and altruism as decisive factors and formulates testable hypotheses for risk preference and altruism. Effects of time preference on food choices have been widely discussed in the literature (Barlow et al., 2016; Smith et al., 2005); in Section 5.2 below, we will also show that particularly time preference is a highly endogenous characteristic that only allows for a limited explanation in regard to nutrition. Therefore, we focus in this empirical chapter on the following two research questions:

1. Does risk preference affect dietary intake and, if so, to what extent?
2. Does altruism affect dietary intake and, if so, to what extent?

The next Section 5.2 presents a global perspective on patience, risk preference and altruism, and links their global distribution with the distribution within the study regions. This section also explains descriptively why time preference is not the focus of the present research. A brief recap of the theoretical model in Section 5.3 clarifies the intention of the empirical analysis.

Assessing economic preferences in an empirical setting presents various practical challenges as consumption behavior does not necessarily reveal these preferences. Time-intensive hypothetical or actual games are played with respondents to elicit them. A survey design that combines preference elicitation with the assessment of relevant indicators for food and nutrition security in larger sample sizes is even more challenging. On the basis of the Global Preference Survey (Falk et al., 2015), we introduce a survey design that overcomes these challenges and that has been successfully implemented in three regions of India. Section 5.4 presents the survey and the collected data. The estimation strategy is based on multiple regression and its limitations are discussed in Section 5.5. The results are presented and the statistical limitations addressed in Section 5.6.

Preferences are naturally linked to economic outcomes. Accordingly, this empirical chapter starts by presenting the spatial distribution of preferences, as well as empirical linkages between preferences and economic behavior in its basic denotation, which is income generation. We consider the global scale and conclude with the study region of this research.

5.2 Distribution of Preferences

Individual preferences only change slowly. Changes can be induced by the general environment, personal situation, or by shocks (Dohmen, Falk, Golsteyn, et al., 2017; Hideki, 2013; Malmendier and Nagel, 2011). However, it is not yet fully explained under which circumstances and to which extent economic preferences vary. The literature tends to regard preferences as relatively stable when considering preferences such as risk affinity, time preference or altruism (Stigler and Becker, 1977; Vermeir and Verbeke, 2006). Accordingly, certain preference-tendencies are often associated colloquially with specific cultures. These associations tend to emerge from stories or personal experiences and do not necessarily reflect true preferences in a representative way. Nevertheless, there are in fact differences in economic preferences across the globe. Falk et al. (2015, 2018; 2016) elicited a set of preferences in 76 countries in 2012 that are representative on country-level and region-level. This allows a comparison of country-specific preferences. Figure 5.1 on the following page presents the levels of patience per country. In this map, green reflects a very high discount rate or respectively the tendency toward immediacy. Red corresponds to the opposite, which is a high level of patience. As the present research uses a similar methodology for eliciting the preferences, it offers the opportunity for a comparison of preferences between the world, India on the whole and the study regions.

It is useful to compare preferences relatively; one person can be more patient than another. Referring to Figure 5.1 on the next page, some nations are more patient than others, but overall the levels of patience are seemingly randomly distributed with a few countries in either extreme. This also holds true for other preferences that are of interest to this research. The global distributions of risk preference and altruism are mapped in the Appendix on page 202. The probability distributions of time, risk and altruism are shown on page 204. The probability distributions indicate that the majority of countries are patient with some being less patient; risk taking is normally distributed and altruism is also normally distributed with a few countries being more altruistic.

128 Effects of Preferences on Food Consumption

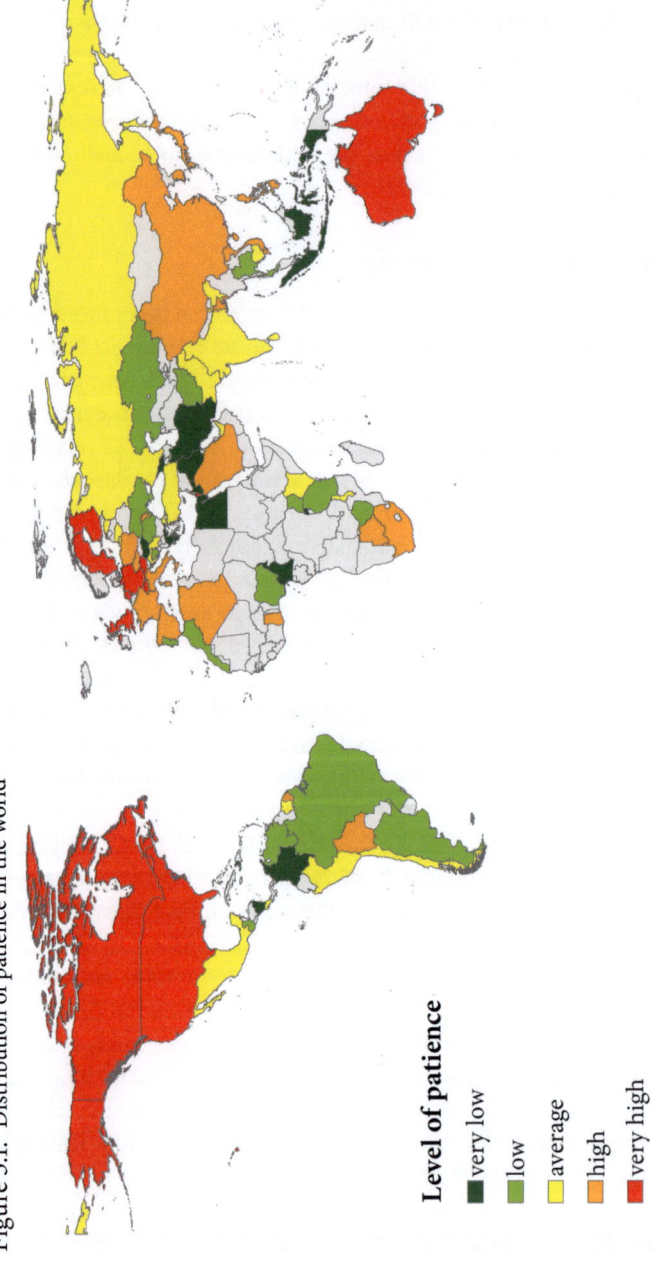

Figure 5.1: Distribution of patience in the world

Data source: Falk et al., 2018; Falk, Becker, Dohmen, Huffman, et al., 2016. Author's illustration

Figure 5.2: Correlation of patience and income

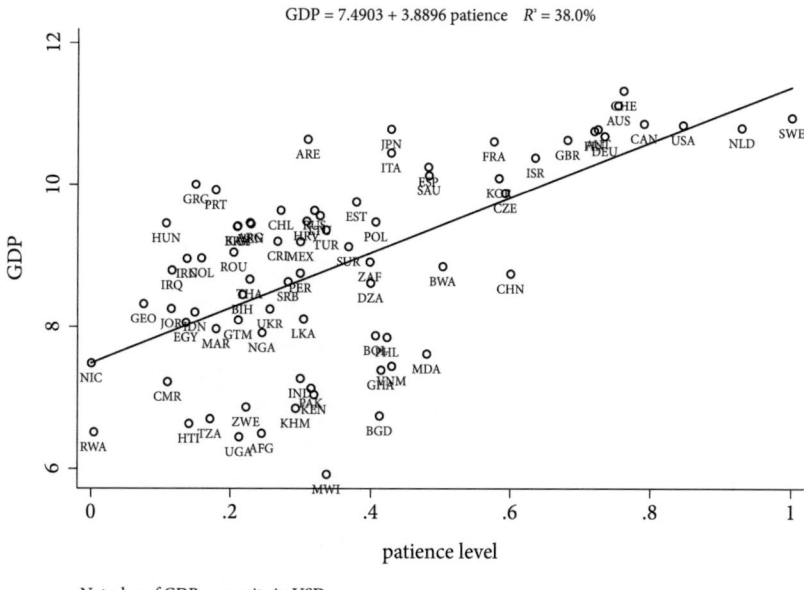

Note: log of GDP per capita in USD

Data source: Falk et al., 2018; Falk, Becker, Dohmen, Huffman, et al., 2016; World Bank, 2017

The literature frequently discusses economic preferences considering income or investment choices. The global preference elicitation presents the opportunity to test this relation globally. Looking again at Figure 5.1 on the previous page, one can notice that the seemingly random distribution of patience is not random; high-income countries tend to be more patient. A basic regression showcases this relationship; see Figure 5.2 above.

The regression indicates a strong association between patience and GDP per capita on a global scale. This association is also observed in a micro-scale economic context. Individuals with low income have few resources available for savings. Consumption usually takes place shortly after a budget is available; hence, the discount rate is high and the level of patience low. The intuition for endogenous preferences is adumbrated thus: the poor individual could be intrinsically less patient, but the circumstances might also force her to act accordingly. On the other hand, risk preference and altruism do not have this strong correlation with income (see Appendix on page 206).

We turn the focus now to India and the study regions, which are the states of Jharkhand, Karnataka and West Bengal. The spatial distribution of the preferences representative for the states are displayed on the maps starting with on page 208 of the Appendix. The population of these three states tend to be patient. Bihar and Jharkhand are quite risk averse whereas Karnataka is highly risk taking; and the three states are among the most altruistic in India. The visual observations of the maps are confirmed when we compare the probability distributions of the preferences between India and the study regions on an individual level (see Appendix starting on page 214). These individual-level distributions also show, although with some deviations in preference tendencies between the study regions and India overall, that the probability distributions have still roughly the same shapes. The correlation between preferences and income does not reflect the associations of the global data. The correlations are slightly positive, but at a small margin, as the R^2 indicate (see Appendix on page 211).

The main take-away from these brief empirical reflections are the following. Firstly, patience is highly correlated with income on global level, but possibly not so in India. Risk preference and altruism do not appear to have a positive association with income. Income is a core factor in food access and nutrition security considerations. Hence, endogeneity of preferences is a valid concern, and given the indication for patience, it is advisable to focus on risk and altruism instead. Conceptual links between risk, altruism and income can still be claimed. Accordingly, robustness checks for various groups need to consider various wealth and income levels as well.

Secondly, the distributions of preferences for the study regions roughly reflect the distributions of preferences for the whole of India. The study regions tend to be more altruistic and patient, but given the probability distributions, conclusions for the study regions on preferences can be generalized for India to a certain extent. The distributions of preferences for the study sample are examined in Section 5.4 after the theoretical links between preferences and nutrition are recapped.

5.3 Theoretical Model

The theoretical framework follows the concept of expected utility maximization in which an individual makes economic decisions about an available budget. The allocation of an available budget can generally be divided into savings and consumption. In an intertemporal allocation decision, savings can mitigate future risks whereas consumption can either satisfy current needs or contribute to a future pay-off (Deaton, 1992). The current and future available budget is in a simple form determined by income, which itself is the product of wage and working hours. Assuming an exogenous wage level, an individual can increase the income by increasing the working hours. This basic life-cycle model forms the basis for an expected utility optimization setting. An individual tries to maximize her utility over time subject to the available budget.

The presented model in Chapter 4 amends the expected utility setting by considering nutrition as an investment. The intuition is rooted in the assumption that food consumption is not only a matter of taste but also a necessity that enables individuals to earn income in the future. The nutrition outcome respective to the health of an individual is her productive asset (according to Grossman, 1972; Levhari and Weiss, 1974). Economic preferences such as patience and risk preference influence the budget allocation decisions. It can be hypothesized that risk preference and time preference have a similar effect direction on savings and on investments in nutrition; risk averse and patient individuals tend to allocate more of their budget to savings and nutrition than to other consumptive goods. A core feature of a model must therefore be to distinguish the effect sizes of preferences. To single out their effects in the current period, the model is first differentiated to find an optimal savings rate. The optimal savings rate is substituted in the FOC over nutrition. The final solution is the optimal level of nutrition that maximizes the expected utility and that is independent from current savings decisions.

The following assumptions are made: (1) An individual makes rational choices to maximize her utility, (2) the nutrition outcome is known and has diminishing returns, (3) income is only generated through activities that are affected by the nutrition outcome, (4) food consumption is an individual decision, and (5) the uncertain future entails normally distributed shocks. Associated limitations are discussed in detail in Section 4.3.4 on page 120.

Following this theoretical approach and solving for the optimal solution yields an optimal nutrition level as:

$$n^* = \frac{\delta^2 \mu^2 w^2}{(A\delta w^2 \sigma^2 + 2p)^2}$$

where δ is the personal discount rate, A is the absolute measure of risk preference[1], μ is the mean and σ^2 is the variance of normally distributed shocks, w is the wage rate, and p is the price of nutrition.

In an extension to the model, altruism is included. Altruism is considered as a mechanism with which someone receives utility through sharing (Becker and

---

[1] Note for the interpretation of the model, an individual is risk averse if $A > 0$, risk neutral if $A = 0$ and risk taking if $A < 0$ (Arrow, 1965; Pratt, 1964).

Barro, 1986; Stark, 1995). The effect of someone's desire to share food with an individual is considered. In this particular setting it is the effect of the husband's altruistic behavior towards his wife by sharing food. The given conceptual link is that the utility of the husband is increased by the factor of his level of altruism if the utility of his wife is increased, where the utility of the wife depends on her nutrition and the additionally received food that has been shared.

The model allows to deduce the hypotheses, which are empirically tested in this chapter.

Hypothesis 1
Risk preference measured as an individual level of risk aversion positively affects the current nutrition level of individuals.

Hypothesis 2
Altruism of another household member j towards an individual i positively affects total utility of i's nutrition due to the increased amount of food items n_i.

5.4 Data

This study relies on primary data that was collected in the states of Jharkhand, West Bengal and Karnataka in India between January and June 2017. Village-level data and market information complement the household-level survey. The sample size of the dataset consists of 1177 households in 111 villages of the 3 regions. The regions were chosen to suite the overall research objective (i.e. rural areas that are prone to environmental shocks such as droughts and floods). We utilized a two-stage cluster sampling technique: the villages were randomly chosen from pre-identified districts, and the households were also randomly chosen, partly on the basis of a random draw from current census data. If the census data was not up to date, we used the random walk technique for identification of the respondent households (see Appendix on page 177). We have individual data for the household head as well as for the spouse of one's household. Nutrition information is available for the female respondent and for the household.

5.4.1 Preference elicitation

Generally, there are two methods for eliciting individual preferences, either through experiments that entail an actual game design or through experiments that use hypothetical games. The intention of both designs is the same, to confront an individual with a cascade system of response options for similarly framed questions. Standard economic choice experiments are most commonly used in economic behavior experiments; however, these require a controlled laboratory setting and are resource intensive. Lab in the field experiments mirror these settings outside of the laboratory and in the communities of interest. The challenges remain, however, and mostly only a fairly small sample size can be surveyed. On the other hand, hypothetical games circumvent these challenges by asking the respondents to imagine an economic game and by inquiring on the hypothetical response to such a game. This procedure is more time-efficient and enables the survey of larger groups. However, hypothetical games are more challenging to the respondent, especially in the field where economic games might present a completely novel concept of interaction. Also, some scholars argue that hypothetical action is different from actual action (Kahneman and Tversky, 1979). Therefore, the choice between actual games and hypothetical games always represents a trade off. Other scholars claim the opposite, namely that there is no difference in the outcome of actual and hypothetical games (Thaler, 1986). Only a few have systematically compared the outcomes, which resulted in mixed results that do not allow for a final answer (e.g. Gillis and Hettler, 2007).

Therefore we are confident in the use of hypothetical games: we have intended to design an interdisciplinary household survey that includes socioeconomic, agricultural and nutritional information as well as for the first time preference elicitations. This survey tool is applicable and reproducible for rural households in marginalized settings and can be applied to large sample sizes with a reasonable budget and in a brief time period. We respond to the criticism regarding hypothetical games by employing an innovative tool for economic games that was designed to elicit preferences at a global scale covering more than 90% of the global population (Falk et al., 2017). The two main criticisms as mentioned above are comprehension difficulties of the respondents and possible differing outcomes in comparison to actual games.

First, to minimize possible comprehension difficulties of the respondents, we further trained the enumerators intensively in the methodology of economic games (both factual and hypothetical). A focus was set on explaining the choice games to a rural population that is not familiar with concepts of behavioral economics.

Second, we utilize the survey preference module that was developed by Falk et al. (2016) and that is based on (Barsky et al., 1997). The survey module "contains the set of items for each preference that best capture revealed preferences in incentivized laboratory experiments, in the sense of an optimal trade-off between explanatory power and parsimony" (Falk, Becker, Dohmen, Huffman, et al., 2016,

Table 5.1: Items of survey preference module

Preferences		Item Description
Risk preference	quantitative	Staircase procedure of 31 hypothetical choices between a lottery and a safe option
	qualitative	How do you see yourself: Are you a person who is generally willing to take risks, or do you try to avoid taking risks?
Time preference	quantitative	Staircase procedure of 31 hypothetical choices between an early payment "today" and a delayed payment "in 12 months"
	qualitative	In comparison to others, are you a person who is generally willing to give up something today in order to benefit from that in the future?
Altruism	quantitative	Today you unexpectedly receive USD 1,600 (INR 2,400). How much of this amount would you donate to a good cause?
	qualitative	How do you assess your willingness to share with others without expecting anything in return when it comes to charity?
Positive reciprocity	quantitative	Willingness to reciprocate by asking for a hypothetical situation: "Which present do you give as a thank-you gift?"
	qualitative	Self-assessment: "When someone does me a favor I am willing to return it"
Negative reciprocity	qualitative	Self-assessment: Willingness to punish if oneself/others treated unfairly
	qualitative	Self-assessment: Willingness to take revenge

Note: Table is adapted for this study from Falk, Becker, Dohmen, Huffman, et al., 2016

p. 10). As such, the module contains for each preference a hypothetical choice experiment and a qualitative item. The first is a quantitatively revealed preference approach and the latter reflects a self-assessment of the respondent. Table 5.1 above stylizes the survey items that compose the various preference elicitations in the survey preference module.

The lottery game that we use for eliciting risk is a five step game, which changes the value of a hypothetical sure payout depending on the previous answer. For example, the respondent is asked if she would prefer a safe payment of INR 320 or instead to play a game at which she has equal chances to either win INR 600 or nothing. Depending on the response, the following safe payment value would

Table 5.2: Scale of preference traits

Preferences	Scale	
Risk	0............................1	
	risk averse	risk loving
Time	0............................1	
	patient	impatient
Altruism	0............................1	
	egoistic	altruistic
Pos. reciprocity	0............................1	
	non-reciprocal	reciprocal
Neg. reciprocity	0............................1	
	non-reciprocal	reciprocal

change to either INR 160 or INR 480. Similarly, the hypothetical game for time preference is played. The staircases for risk and time preference are displayed in the Appendix on page 218 and on page 219. After five turns, the respondent will have revealed a value at which she is factually indifferent. Given the set-up of the game, there are 32 possible indifference values.

Whereas the survey reveals the preferences for risk, time, trust, altruism, negative and positive reciprocity, we focus in this study primarily on risk and altruism. However, the analysis will utilize the other preferences either for robustness checks or for gaining additional insights. A complete set of the English version[2] of the survey preference module as we used it in the Indian setting can be found in the Appendix on page 217.

Individual-level indices were computed for each of the preference traits on the basis of Falk et al. (2017). Each index was created by calculating the z-scores of each response item on the individual level and by weighing the z-scores; the weights are the result of the validation study (Falk, Becker, Dohmen, Huffman,

---

[2] The surveys were translated in Bengali, Hindi and Kannada for the use in the three different study regions, the translated versions are not provided.

et al., 2016). The computation is explained in the Appendix on page 220. We modified the final index by normalizing each trait for gaining a continuous scale from 0 to 1 (see Table 5.2 on the previous page).

The preference traits are revealed for each participating household in the survey, separately for the household head and the spouse[3]. In case the household head or spouse are not available (e.g. due to temporary migration or deceased), we only elicit the preferences of the present respondent. Z-scores for the indices are generated on the complete sample including household heads and spouses.

For the average household head we can state the following: he tends to be risk averse, is more altruistic than egoistic, has a relatively high discount rate, and tends to act more positively reciprocal than negatively reciprocal. For the average spouse, the same characteristics hold true; however, women tend to be even more risk averse than men. On average, the preference traits of the household head and spouse tend to be strongly correlated within households (we are comparing the average of the full sample with few changes in response numbers due to infrequent skipped responses, see Table 5.3 below).

The full sample averages give a broad understanding of the distribution of preference traits. Examining the main preferences that are of interest to this study seems necessary to recognize possible intra-household differences as well as the distribution of risk and altruism levels. Figure 5.3 on the following page shows the distribution of risk levels as probability density estimates per household head and

Table 5.3: Preference traits statistics

	Household head			Spouse			Correlation
	N	Mean	SD	N	Mean	SD	
Risk	1177	0.43	0.26	1177	0.38	0.25	0.46
Altruism	1177	0.54	0.18	1175	0.52	0.19	0.52
Time	1176	0.36	0.26	1174	0.36	0.26	0.46
Pos. Reciprocity	1157	0.63	0.21	1147	0.61	0.20	0.64
Neg. Reciprocity	1177	0.41	0.26	1175	0.41	0.26	0.78

---

3   In the sample only 0.01% of all households are female headed, therefore we will use interchangeably the terminology female for spouse and male for household head in the following.

Figure 5.3: Distribution of risk among gender

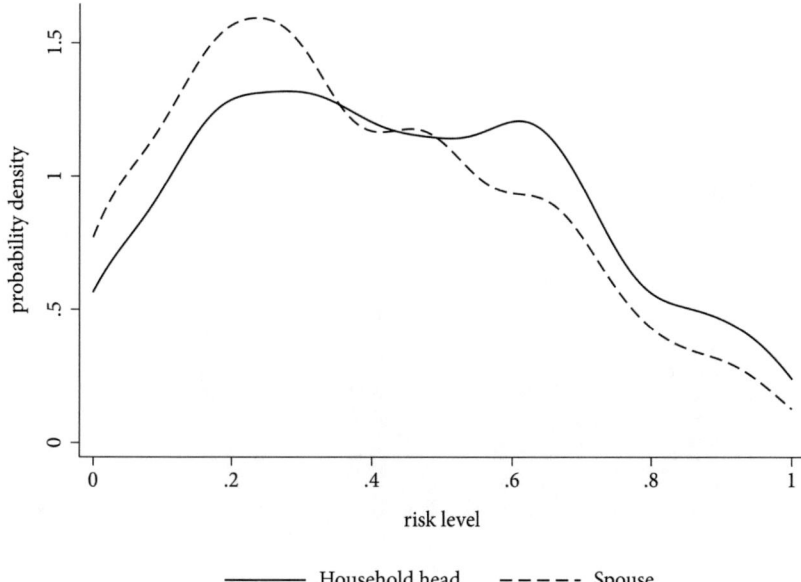

Note: The distribution is computed as kernel density estimation using a Gaussian kernel

spouse. On average, the majority of individuals is risk averse and women tend to be more risk averse than men.

Figure 5.4 on the following page displays the risk levels per household. The five bins represent quintiles of risk level distribution with the first bin being risk averse, the third bin relatively risk neutral and the fifth bin risk taking. There are five groups of distributions, each group representing the bin for the risk level of the household head. Within each group, the distribution of risk levels per spouse is shown. It is apparent that on the basis of the quintiles, the household head and spouse tend to have the same risk level. This is a noteworthy observation, as it might hint at homogeneously characterized households when a household would only be comprised of household head and spouse (compare Kimball et al., 2009). The correlation coefficients between the means of the risk levels by gender in Table 5.3 on page 136 hint at a similar finding. However, the variation of risk levels on individual level is still big enough when considering the exact values (and no bins) to reject a homogeneous behavior.

Figure 5.4: Risk levels by quintiles among households

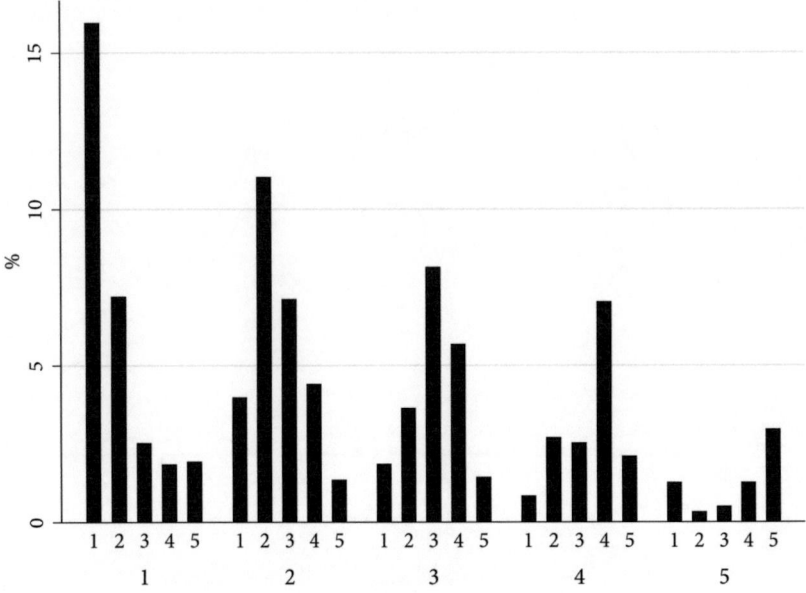

Figure 5.5: Distribution of altruism among gender

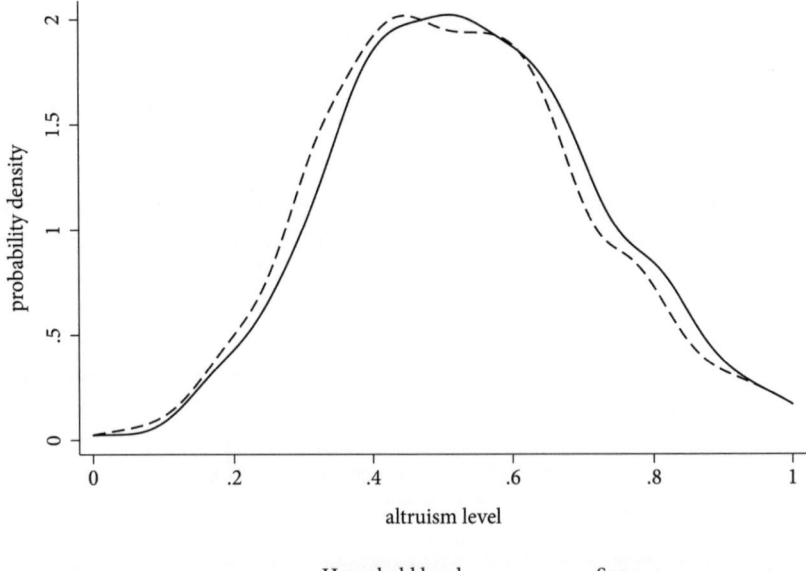

Note: The distribution is computed as kernel density estimation using a Gaussian kernel

Figure 5.6: Altruism levels by quintiles among households

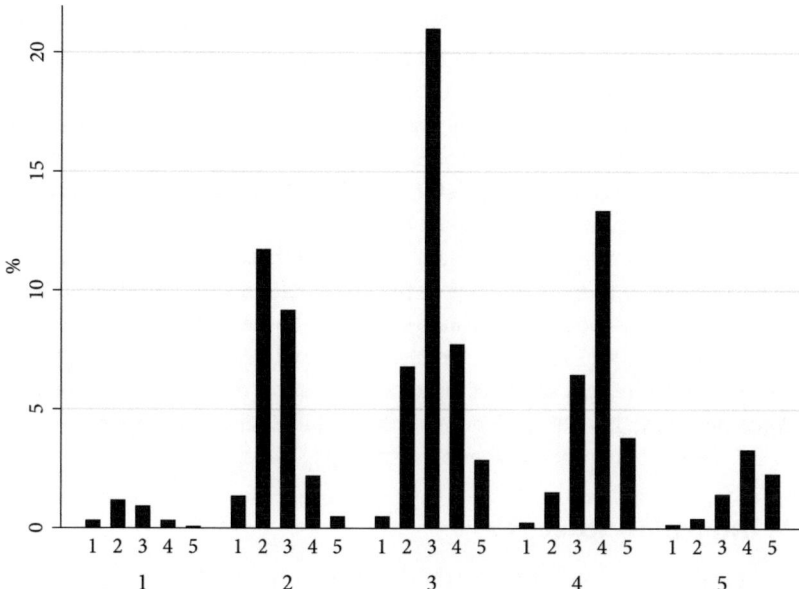

Figure 5.5 on the previous page shows the distribution of altruism levels as kernel density estimate per household head and spouse. The distribution of altruism is in both genders normally distributed. There are no significant differences between the altruism levels per gender and we see a slight negative skew.

Figure 5.6 above groups the altruism levels per household. Aside from the normal distribution, we see more credibility for a slightly more altruistic than egoistic behavior of the individuals. And again, similar to the risk levels, the preference levels within the household tend to be roughly in line among household head and spouse. The distributions for time preference, positive and negative reciprocity show a similar characteristic as for risk and altruism. These can be found in the Appendix on page 221. With a better understanding on the preference levels of the population, we will describe the basic statistics of the main variables for the empirical analysis in the next Section.

5.4.2 Main variables

Our main dependent variable is the Dietary Diversity of Women, which we have explained in detail in Section 3.4.1 on page 70. For robustness checks, we use the Household Dietary Diversity Score and the Food Insecurity Experience Scale. Figures of their distributions can be found in the Appendix on page 224 and on page 225. We use a wide set of control variables that we categorize into individual-level, household-level and village-level variables. The variables that are used in this study are described in Table 5.4 on page 143.

The average household has 4.8 members, is male-headed and has a gender ratio of 0.49 women to men. The age of the household head is approximately 39 years, the age of the spouse is 31 years. The average household head has 5 years of formal education and 79% of the sample's household heads are working primarily in agriculture-related activities either as a farmer or day laborer. The average income per capita is INR 1,385 per month at adult equivalence (approximately USD 23). Table 5.5 on page 145 shows the summary statistics in detail.

5.5 Estimation Strategy

The dependent variable DDW can take only non-negative integer values as these present count data: 0,1,2,...,10 . Further, from the descriptive statistics we can see that the probability distribution of DDW resembles a Poisson distribution (see Figure D.25 in the Appendix on page 224). Accordingly, we use a Poisson regression model with the following basic specification:

$$DDW_i = exp\left\{\beta_0 + \beta_1 Risk_i + \beta_2 Altruism_j + \beta_3 X_{ihvr}\right\} + \epsilon_{ijhvr},$$

with r = 1,...,R, v = 1,...,V_r, h = 1,...,H_{vr}, i = 1,...,I_{hvr} and j = 1,...,J_{hvr}. X_{ihvr} is a vector of explanatory variables and a set of control variables on village, household and individual level, β_3 is a vector with corresponding regression coefficients and ϵ_{ijhvr} is the error term. Of particular interest are the coefficients β_1 for $Risk_i$ representing the risk level of the spouse and β_2 for $Altruism_j$ representing the altruism level of the household head.

In the following, we will also perform the analysis by using an OLS regression for secondary results and for robustness checks. It differs from the Poisson regression in that we standardize the dependent variable:

$$log(DDW_i) = \beta_0 + \beta_1 Risk_i + \beta_2 Altruism_j + \beta_3 X_{ihvr} + \epsilon_{ijhvr}$$

Following from the theoretical model 4.9 on page 120, we can expect to see positive correlations between $Risk_i$ and DDW as well as between $Altruism_j$

Table 5.4: Description of variables

Variables	Description
Individual preferences	
Risk level of spouse	Risk level revealed from spouse, measured from 0 to 1
Risk level of household head	Risk level revealed from household head, measured from 0 to 1
Altruism of spouse	Altruism revealed from spouse, measured from 0 to 1
Altruism of household head	Altruism revealed from household head, measured from 0 to 1
Patience of spouse	Patience revealed from spouse, measured from 0 to 1
Patience of household head	Patience revealed from household head, measured from 0 to 1
Pos. reciprocity of spouse	Positive reciprocity revealed from spouse, measured from 0 to 1
Pos. reciprocity of household head	Positive reciprocity revealed from household head, measured from 0 to 1
Neg. reciprocity of spouse	Negative reciprocity revealed from spouse, measured from 0 to 1
Neg. reciprocity of household head	Negative reciprocity revealed from household head, measured from 0 to 1
Nutrition variables	
Dietary Diversity Score	Number of food groups consumed by woman in past 24 hours on a scale of 9 food groups
Minimum Dietary Diversity Score	At least 5 food groups are consumed in past 24 hours on scale of 10 food groups (1 = yes, 0 = no)
Household Dietary Diversity Score	Number of food groups consumed by household in past 7 days on a scale of 12 food groups
FIES	Food Insecurity Experience Scale on a scale from 0 to 8
Height of spouse	Height of spouse in cm
Weight of spouse	Weight of spouse in cm
BMI of spouse	Body Mass Index of spouse
Individual variables	
Age in years	Age of spouse in years
Literacy of spouse	Literacy level of spouse (0 = illiterate, 1 = literate)
Formal education of spouse	Years of formal education of spouse

Table 5.4: *(continued)*

Variables	Description
Household variables	
Age of household head	Age of household head in years
Literacy of household head	Literacy level of household head (0 = illiterate, 1 = literate)
Formal education of household head	Years of formal education of household head
Female headed household	Binary variable if household is headed by female (1 = yes, 0 = no)
Household members	Number of household members
Religion: Hindu	Binary variable if religion is Hindu (0 = no, 1 = yes)
Religion: Muslim	Binary variable if religion is Muslim (0 = no, 1 = yes)
(log) Income	Monthly income on adult equivalence scale measured in expenditures
(log) Total value of liquidable assets	Value of liquidable assets
Total land size	Total land available for agricultural production in hectares
Non-farm occupation	Binary variable if primary occupation is non-farm (0 = no, 1 = yes)
Number of government schemes	Count variable of number of government schemes that the household was utilizing (counted if at least one member is using a scheme)
Distance to next market in km	Distance to next market in km
Household is electrified	Binary variable if household has electricity (0 = no, 1 = yes)
Regular market visits	Binary variable if a member of the household is visiting the next available market regularly (1= yes, 0 = no)
Village variables	
(log) Village population	Population size of the household's village
(log) Village population 10 years ago	Population size of the household's village 10 years ago
Years that village is electrified	Number of years since the household's village received access to electricity
Women group	% of women in village that participate in at least one women group
Women group 10 years ago	% of women in village that participate in at least one women group 10 years ago
NGO support	Count variable of number of NGO programs that are active in the village

Table 5.4: *(continued)*

Variables	Description
NGO support 10 years ago	Count variable of number of NGO programs that are active in the village 10 years ago
Village infrastructure	5-point Likert scale if village is well connected by road (1 = very bad to 2 == very good)
Village infrastructure 10 years ago	5-point Likert scale if village was well connected by road 10 years ago (1 = very bad to 2 == very good)
Years to last covariate shock	Number of years since the village suffered from the last covariate shock
Day labor employment situation	5-point Likert scale if day labor employment is easy (1 = very bad to 2 == very good)
Day labor employment situation 10 years ago	5-point Likert scale if day labor employment was easy 10 years ago (1 = very bad to 2 == very good)
Regions	
Jharkhand	Binary variable if household lives in region Jharkhand (0 = no, 1 = yes)
Karnataka	Binary variable if household lives in region Karnataka (0 = no, 1 = yes)
West Bengal	Binary variable if household lives in region West Bengal (0 = no, 1 = yes)

and DDW[4] From a statistical point of view, however, various issues might limit the interpretation of the coefficients. We cannot exclude the possibility that preferences are endogenous to possible unobserved variables. Strictly speaking, we cannot even exclude that the preferences are endogenous to some included control variables. Theoretical and other empirical literature indicate that preferences are intrinsic to individuals, although preferences as well as all human behavior and characteristics are partly formed by external factors. Its thus follows that,

---

4   A second expectation that follows from the model is to have a nonlinear relation between risk and nutrition. For reasons of simplicity, we assume a linear relationship at this point. The distribution of risk levels in Figure 5.3 on page 137 indicates that the majority of individuals have a risk level of 0.75 or less, and respectively tend to be risk averse. Comparing with the model prediction for the relation between risk and nutrition in Figure 4.2 on page 118, we see that for individuals with an Arrow-Pratt measure of $-0.5 < A \leq 1$ (which relates to < 0.75 in the risk level scale that is used here) the relation between risk and nutrition is sufficiently linear.

Table 5.5: Summary statistics for variables

Variables	Mean	Median	SD	Min	Max
Individual preferences					
Risk level of spouse	0.38	0.33	0.25	0	1
Risk level of household head	0.43	0.42	0.26	0	1
Altruism of spouse	0.52	0.50	0.19	0	1
Altruism of household head	0.54	0.53	0.18	0	1
Patience of spouse	0.36	0.31	0.26	0	1
Patience of household head	0.36	0.31	0.26	0	1
Pos. reciprocity of spouse	0.61	0.61	0.20	0.06	1
Pos. reciprocity of household head	0.63	0.64	0.21	0	1
Nutrition variables					
Dietary Diversity Score	3.57	3	1.17	1	7
Minimum Dietary Diversity Score	0.24	0	0.43	0	1
Household Dietary Diversity Score	6.78	7	1.35	3	11
FIES	2.93	2	2.81	0	8
Height of spouse	153.21	153	6.76	126	175
Weight of spouse	51.46	51	9.24	33	87
BMI of spouse	21.86	21.56	3.35	14.86	35.71
Individual variables					
Age of spouse	30.59	26	11.82	15	80
Literacy of spouse	0.62	1	0.48	0	1
Formal education of spouse	5.82	7	4.23	0	17
Household variables					
Age of household head	39.15	35	13.62	19	90
Literacy of household head	0.57	1	0.50	0	1
Formal education of household head	5.36	5	4.36	0	17
Female headed household[a]	0.01	0	0.12	0	1
Household members	4.81	4	1.62	2	14
Religion: Hindu	0.67	1	0.47	0	1

Table 5.5: *(continued)*

Variables	Mean	Median	SD	Min	Max
Religion: Muslim	0.33	0	0.47	0	1
(log) Income	6.94	6.86	0.72	4.63	9.48
(log) Total value of liquidable assets	10.35	10.35	1.99	5.70	15.58
Total land size	0.88	0.61	1.27	0.01	24.58
Non-farm occupation	0.21	0	0.41	0	1
Number of government schemes	3.25	3	1.20	0	7
Distance to next market in km	3.92	2	4.23	0.07	20
Household is electrified	0.77	1	0.42	0	1
Regular market visits	0.70	1	0.46	0	1
Village variables					
(log) Village population	7.19	7.09	1.11	4.32	9.39
(log) Village population 10 years ago	6.83	6.62	1.08	4.09	9.21
Years that village is electrified	22.47	9	20.35	0	67
Women group	64.49	70	26.33	5	100
Women group 10 years ago	23.52	20	22.45	0	100
NGO support	2.57	3	1.06	1	4
NGO support 10 years ago	1.88	2	0.75	1	4
Village infrastructure	3.06	3	1.12	1	5
Village infrastructure 10 years ago	2.02	2	0.73	1	3
Years to last covariate shock	15.96	17	12.57	1	57
Day labor situation	2.76	3	1.09	1	5
Day labor situation 10 years ago	2.32	2	0.82	1	4
Regions					
Jharkhand	0.29	0	0.45	0	1
Karnataka	0.37	0	0.48	0	1
West Bengal	0.34	0	0.47	0	1

Sample size is 1177 households

[a] Only 15 households are female headed

firstly, we will check for multicollinearity in our analysis. Even if statistical checks claim a low multicollinearity, which is methodologically irrelevant, we might potentially overestimate the effects of preferences on the dependent variables or even face type 1 errors in case of low significance levels of the computed coefficients. *Secondly*, from a theoretical point of view, reverse causality could be an issue. Good nutrition over a long period of time increases the productivity of individuals and on average their income. One could argue that increased income and the physical well-being of good nutrition possibly also change risk taking behavior, altruism and other preference levels. Inversely, bad nutrition over a long period might force individuals to be cautious with future nutrition compared to acting in a risk averse way. These time-related effects cannot be estimated with the proposed model nor with the available data set. However, we are confident that the effects for our sample will hold the causality claims given our focus on an average effect at one point in time. To respond to possible time-effects within our limited scope, we will also check if certain groups that might have consumed a particularly good or bad diet for a longer period (assumingly the top or the lowest wealth quintiles) result in different estimates. Ultimately, we will not be able to fully incorporate a perfect identification strategy due to the nature of preferences. This is also partly caused by missing unobservable characteristics that have been discussed in the literature. Namely, the ability to make rational decisions and long-term decision making quality (Burks et al., 2009; Choi et al., 2014). These can potentially affect economic outcomes in interaction with economic preferences, hence, affect nutrition choices (Castillo et al., 2018; Jacobson and Petrie, 2009). Accordingly, causality claims on the basis of the presented empirical analysis are not possible. To satisfy a rigorous analysis on the basis of correlations, we will elaborate extensively the sensitivity of the results and include sufficient robustness checks. The aim is to test the robustness of the results in many different ways in order to present correlations that approximate to causal claims. Let us first consider the results of the regressions before we put these into perspective with the robustness checks.

5.6 Results

5.6.1 Primary results

The primary objective is to estimate the correlation coefficients of risk levels from the female and the altruism levels of the male with dietary diversity of the spouse. Table 5.6 on page 148 presents the econometric results. The table displays four regressions with OLS and four regressions with Poisson specification. The regressions vary by the included covariates, which are clustered on the individual, household and village levels. Independent of the covariates and the specification used, risk preference and altruism have a highly significant effect on dietary diversity. The significance levels are only slightly reduced to the 0.05 level when all covariates are included. Regarding the coefficients, the relevance of each coefficient is reduced in the full model with all covariates (7 and 8) in comparison to the simple correlation without covariates (1 and 2). Considering the Poisson specifications, the effect size of risk preference on dietary diversity is between 8.7% and 13.5% per 1 unit increase of risk level, and the effect size of altruism on dietary diversity is between 11.4% and 30.0% per 1 unit increase of altruism level[5]. Given the scale of preference levels (0 to 1), it is more reasonable to interpret the results with 10 percentage points increments. Thus, risk preference has a rounded effect size between 0.9% and 1.4% and altruism has an effect size between 1.1% and 3.0%. The interpretation of the results can only be indicative and not literal, but the tendency occurs that altruism of the household head is more relevant than the risk level of the spouse.

We need to take a closer look at the results for the covariates to confirm that the model is specified in line with theory and expectations and that it does not produce results by chance. Table 5.7 on page 150 presents the results of the full model in OLS and Poisson specification while also displaying the coefficients for all covariates. There are slight differences in the significance levels as well as in the coefficients between OLS and Poisson; we consider both for the interpretation. Significantly and positively correlated to dietary diversity are the literacy level of the female, income, wealth, religion, larger villages, electrification in villages, and participation in women's groups. Slightly significantly and negatively correlated are the number of males and females below 5 years of age in the household and NGO support. Aside from the negative effect of NGO support, the results are generally not surprising and lend credibility to the outcome that risk preference of the spouse and the altruism of the household head are highly significantly (at 0.05 level) correlated to the spouse's dietary diversity.

The different units for the variables prohibit a direct comparison of the coefficients' relevance. Nevertheless, we would like to showcase some effects for comparative purposes and use the Poisson specification for this. The size of the

---

5   For interpreting the coefficients of the Poisson specifications, we calculate the IRR of each coefficient. Considering the coefficient of risk level in specification (2) then $exp(0.127)$ results to 1.135, which is a 1.35% increase if the independent variable increases by 1 unit.

Table 5.6: Regression of spouse's risk levels and household head's altruism levels on dietary diversity

Dependent variable	(1) OLS (log) DDW	(2) Poisson DDW	(3) OLS (log) DDW	(4) Poisson DDW	(5) OLS (log) DDW	(6) Poisson DDW	(7) OLS (log) DDW	(8) Poisson DDW
Risk level of spouse	0.122***	0.127***	0.095**	0.103***	0.083**	0.085**	0.085**	0.084**
	(0.041)	(0.037)	(0.039)	(0.035)	(0.039)	(0.034)	(0.041)	(0.035)
Altruism of household head	0.314***	0.262***	0.268***	0.220***	0.172***	0.119**	0.162***	0.108**
	(0.057)	(0.051)	(0.056)	(0.051)	(0.060)	(0.053)	(0.061)	(0.053)
Individual covariates	no	no	yes	yes	yes	yes	yes	yes
Household covariates	no	no	no	no	yes	yes	yes	yes
Village covariates	no	no	no	no	no	no	yes	yes
Observations	1177	1177	1165	1165	1137	1137	1102	1102
Adjusted R^2	0.034		0.107		0.174		0.194	
Pearson goodness-of-fit		434.69		393.95		343.97		324.20
p-value		1.00		1.00		1.00		1.00

Robust standard errors clustered by household in parentheses
*** $p<0.01$, ** $p<0.05$, * $p<0.1$

Table 5.7: Regression of spouse's risk levels and household head's altruism levels on dietary diversity, detailed

Dependent variable	(1) OLS (log) DDW		(2) Poisson DDW	
Individual preferences				
Risk level of spouse	0.085**	(0.041)	0.084**	(0.035)
Altruism of household head	0.162***	(0.061)	0.108**	(0.053)
Individual variables				
Age of spouse	0.000	(0.001)	0.001	(0.001)
Literacy of spouse	0.085***	(0.027)	0.067***	(0.023)
Household variables				
Age of household head	0.002*	(0.001)	0.002*	(0.001)
Literacy of household head	0.007	(0.024)	0.016	(0.021)
Female headed household	−0.045	(0.070)	−0.072	(0.069)
Religion: Muslim	0.054	(0.033)	0.053*	(0.032)
Number of males 0-5 years	−0.033*	(0.019)	−0.028	(0.017)
Number of males 5-15 years	0.024	(0.017)	0.013	(0.015)
Number of males 15-60 years	−0.017	(0.017)	−0.018	(0.014)
Number of males 60+ years	−0.014	(0.034)	−0.013	(0.028)
Number of females 0-5 years	−0.036*	(0.020)	−0.032*	(0.017)
Number of females 5-15 years	−0.004	(0.016)	−0.005	(0.013)
Number of females 15-60 years	0.017	(0.017)	0.013	(0.015)
Number of females 60+ years	−0.017	(0.034)	−0.002	(0.028)
(log) Income	0.072***	(0.021)	0.075***	(0.018)
(log) Total value of liquidable assets	0.022***	(0.008)	0.022***	(0.007)
Non-farm occupation	−0.041	(0.030)	−0.027	(0.027)
Regular market visits	−0.025	(0.023)	−0.021	(0.020)
Number of government schemes	0.010	(0.009)	0.007	(0.008)

Table 5.7: *(continued)*

	(1) OLS		(2) Poisson	
Dependent variable	(log) DDW		DDW	
Village variables				
(log) Village population	0.035**	(0.014)	0.032***	(0.012)
Years that village is electrified	0.001	(0.001)	0.002*	(0.001)
Village infrastructure	0.015	(0.012)	0.014	(0.010)
Day labor employment situation	0.006	(0.013)	0.002	(0.012)
Women group	0.001*	(0.000)	0.001*	(0.000)
NGO support	−0.025	(0.015)	−0.027*	(0.015)
Years to last covariate shock	0.001	(0.001)	0.000	(0.001)
Observations	1102		1102	
Adjusted R^2	0.194			
Pearson goodness-of-fit			324.20	
p-value			1.00	

Robust standard errors clustered by household in parentheses

* $p<0.10$, ** $p<0.05$, *** $p<0.01$

household also affects the dietary diversity. Having one child more of either gender and up to 5 years old reduces the dietary diversity score of women by up to 3.2%[6]. The causality is likely caused by the need to nourish the child and the need for a more intense care by the mother. This reduces the total available food for the woman, or potentially the individual preference shifts from nourishing herself to nourishing her child.

The literacy level in combination with education of the woman is an often discussed indicator in development literature for economic improvements such as income generating activities, but also for gender-related empowerment purposes. We chose the literacy level primarily because the population we are studying is

---

6   Calculating the IRR: $exp(-0.032)$ results to 0.9685, hence, to 3.15% decrease.

largely marginalized in rural areas, where the formal education level is generally low and quite homogeneous. The literacy level on the other hand displays a characteristic that can effectively affect the relative status among the population in terms of income, assets or - as in our case - nutrition. The highly significant results for this particular variable confirm this reasoning. We include literacy as a binary variable; a literate woman has a dietary diversity score that is 6.9% higher than the score of an illiterate woman[7].

Finally, we consider income. Everything else held constant, the dietary diversity score increases by 7.8% if (log) income is increased by 1 unit[8], keeping in mind that income is measured as total income of the household in adult equivalent scale. These effect sizes of observable characteristics put the coefficients of risk preference and altruism in perspective. Risk level and altruism have nominally smaller effects. However, given the revelation of presumably unobservable preferences, we can reveal a sizable effect on the nutrition status purely by a different personal tendency for risk and altruism respectively without changing any socioeconomic conditions. We now turn our focus to various nutrition indicators and other secondary results.

5.6.2 Secondary results

Effect on other nutrition indicators

Having a strong and significant effect on nutrition intake might be caused by the methodology for calculating the dietary diversity score for women (DDW) as it was used in the regressions. Estimating the effects on other nutrition intake measures can give the results more credence; additionally, we get a better view of the effects of preferences on various dimensions of food and nutrition security. In the choice of additional nutritional intake measures, we are limited by those measures that are directly influenced by the woman, which are in this case the HDDS and the FIES. We discussed both measures in detail in Section 1.5 on page 38. The HDDS is contrary to its name (Household Dietary Diversity Score) a food access measure that counts the number of food groups consumed by the household in the past 7 days on a basis of twelve different food groups; the higher the score, the better the food access dimension. The FIES is a subjective measure;

---

7   Calculating the IRR: $exp(0.067)$ results to 1.0693, hence, to 6.93% increase.
8   Calculating the IRR: $exp(0.075)$ results to 1.0779, hence, to 7.79% increase.

it asks the respondent eight binary questions (1 = yes, 0 = no) on foods for the past twelve months; the lower the score, the better the food security situation in the past year. Table 5.8 on the following page displays the results for six Poisson regressions with the dependent variables DDW, HDDS, and FIES, each without controls and with controls. Poisson is the methodologically correct specification for these count variables as dependent variables and considering their probability distribution.

In Table 5.8 the models (1) and (2) are known from the previous Section. Models (3) and (4) show the results for the HDDS. Risk and altruism are both highly significantly correlated, though the effect sizes are smaller for this household measure compared to the individual nutrition intake measure. This should not be surprising, since it can be assumed that the effect is directed "via the woman" to the total household, and the woman represents on average a fifth of all household members.

Although having negative effect sizes in models (5) and (6), these coefficients also confirm the previous findings. The negative coefficients indicate a reduction of the subjectively experienced food insecurity, hence, an improvement of food intake. The Pearson goodness-of-fit tests also indicate that (5) and (6) might not be fully correct, so these last results should be interpreted carefully. Still, these secondary results tend to confirm the primary results of the previous section and add to the understanding of the relevance of preferences to food and nutrition security.

Other preferences

Our initial intuition about the research question also involved other preference measures that are widely used in behavioral economics, although which are of no direct relevance in the theoretical model of this study. Positive reciprocity is often confused with altruism because the direction of the effect would be the same. In estimating the effect, it should not matter if an individual acts altruistically out of charity or because the individual expects a positive reciprocal behavior in the future. However, the difference is of relevance when we consider the utility maximization approach that was introduced earlier. Altruistic behavior is rewarded in the same time period since the personal utility is a function of the other's utility, whereas reciprocity is usually rewarded in a future time period, which might not be as useful to an individual who is discounting the future. Looking at the current

Table 5.8: Correlation of risk levels with food and nutrition security indicators

	(1)	(2)	(3)	(4)	(5)	(6)	(7)	(8)
	Poisson	Poisson	Poisson	Poisson	Poisson	Poisson	Poisson	Poisson
Dependent variable	DDW, 9	DDW, 9	DDW, 10	DDW, 10	HDDS	HDDS	FIES	FIES
Risk level of spouse	0.127***	0.084**	0.197***	0.131***	0.109***	0.073***	−0.476***	−0.269**
	(0.037)	(0.035)	(0.043)	(0.038)	(0.025)	(0.025)	(0.131)	(0.117)
Altruism of household head	0.262***	0.108**	0.314***	0.132**	0.115***	0.100***	−0.478***	−0.257*
	(0.051)	(0.053)	(0.057)	(0.056)	(0.033)	(0.035)	(0.161)	(0.155)
Individual covariates	no	yes	no	yes	no	yes	no	yes
Household covariates	no	yes	no	yes	no	yes	no	yes
Village covariates	no	yes	no	yes	no	yes	no	yes
Observations	1177	1102	1177	1102	1177	1102	1174	1099
Pearson goodness-of-fit	434.69	324.20	578.13	393.91	306.70	256.49	3179.70	2838.75
p-value	1.00	1.00	1.00	1.00	1.00	1.00	1.00	1.00

Robust standard errors clustered by household in parentheses
*** $p<0.01$, ** $p<0.05$, * $p<0.1$

period, which the available dataset essentially does, we can expect a smaller effect of positive reciprocity than altruism.

The second preference that is widely discussed is discounting respectively the patience of individuals. Table 5.9 below shows four models: (1) is the initial regression set up that was defined before, (2) replaces altruism levels of the household head with positive reciprocity levels of the household head, (3) includes patience of the spouse to the initial regression, and (4) includes patience of the spouse and replaces altruism with positive reciprocity. All regressions include all previously defined covariates.

The coefficients indicate that positive reciprocity of the household head is not significantly correlated with dietary diversity of the spouse, but it is positively

Table 5.9: Correlation of various preferences with nutrition intake variables

	(1) Poisson	(2) Poisson	(3) Poisson	(4) Poisson
Dependent variable	DDW	DDW	DDW	DDW
Risk level of spouse	0.102***	0.117***	0.094**	0.110***
	(0.034)	(0.034)	(0.037)	(0.036)
Altruism of household head	0.090*		0.092*	
	(0.053)		(0.053)	
Pos. reciprocity of household head		0.048		0.050
		(0.043)		(0.044)
Patience of spouse			0.029	0.025
			(0.038)	(0.037)
Individual covariates	yes	yes	yes	yes
Household covariates	yes	yes	yes	yes
Village covariates	yes	yes	yes	yes
Observations	1102	1082	1099	1079
Pearson goodness-of-fit	307.10	293.16	306.28	292.31
p-value	1.00	1.00	1.00	1.00

Robust standard errors clustered by household in parentheses

* $p<0.10$, ** $p<0.05$, *** $p<0.01$

associated with it. In any case, it has a smaller effect to altruism but has the same direction as the previous brief discussion mentioned. On the other hand, patience of the spouse is not correlated at all with her nutrition. A finding that is in line with other economic literature (Epstein et al., 2014; Rieger, 2015), but still unintuitive given our theoretical reasoning.

Income elasticities

We discussed the links between time preference and income earlier. Given the endogeneity concerns of the other preferences, we discuss the possible link between risk preference and income at this point. The literature generally discusses the influence of risk preference in regard to income generation, investments and innovation (Chetty and Szeidl, 2007; Gatzweiler and von Braun, 2016; Holt and Laury, 2002; Liu, 2013). Along the line, the dataset allows us to refer to point elasticities in order to understand the sensitivity between risk levels and income in India. Table 5.10 displays the point elasticities disaggregated per wealth quintile.

The wealth quintiles are generated on the basis of the total value of liquidable assets per household for the full sample of 1177 households (similar to the regression variable for assets, see 5.4 on page 143). The underlying regression is the main OLS regression that is displayed in Table 5.7 on page 149. Because the main income generating activity is usually performed by the household head, there is a direct linkage between the household head's preference level and income, whereas the correlation between the spouse's preference level and income is only of informative character. As we have seen in the distribution of preferences above, preference levels among the two household members are closely aligned.

Table 5.10: Point elasticities of risk on income

		Risk levels	
		hhh	spouse
Wealth	1	0.078	0.137
quintiles	2	0.086	0.056
	3	0.041	0.106
	4	0.111	0.077
	5	-0.181	0.194
at means		**.081**	**.095**

Figure 5.7: Point elasticities of risk on income

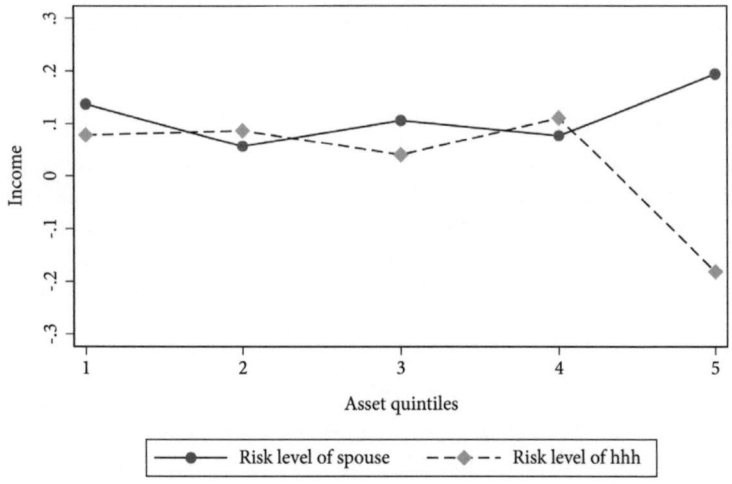

For comparative reasons we therefore display the elasticities of the risk levels of the household head as well as of the spouse.

The point elasticities of the household head vary from -0.181 to 0.111 and of the spouse from 0.056 to 0.194. Aside the fifth wealth quintile, the estimated elasticities seem relatively stable; the large deviance in the fifth quintile might be due to statistical noise. Therefore, the point elasticities at means give a more accurate understanding for the average elasticity (calculated not per wealth quintile, but using the mean value for each included variable of the regression). The positive elasticity indicates that individuals with more risky behavior tend to have a higher income, though the effect size is rather small: a 1% increase in risk level, contributes to a 0.08% higher income. Figure 5.7 above visualizes the relationship graphically.

Decomposing risk preference

We would like to decompose the risk preference measure for understanding what drives the risk assessment of individuals. As explained in Section 5.4.1 on page 132, the preferences are composed by one quantitative hypothetical game and by one subjective self-evaluation. Table 5.11 on the following page presents the decomposition results and helps us to decide if the results are driven by a specific component. The results suggest that the self-assessment is driving the results

Table 5.11: Decomposing risk preference

	(1) OLS	(2) OLS	(3) OLS	(4) OLS
Dependent variable	(log) DDW	(log) DDW	(log) DDW	(log) DDW
Risk level of spouse	0.085**			
	(0.041)			
Z-values of game outcome		0.011		0.006
		(0.010)		(0.011)
Z-values of subjective assessment			0.022**	0.021**
			(0.010)	(0.010)
Individual covariates	yes	yes	yes	yes
Household covariates	yes	yes	yes	yes
Village covariates	yes	yes	yes	yes
Observations	1102	1102	1102	1102
Adjusted R^2	0.194	0.192	0.194	0.194

Robust standard errors clustered by household in parentheses
* $p<0.10$, ** $p<0.05$, *** $p<0.01$

because only the subjective measure is significant and also has a bigger effect size. The weighting puts a bit more emphasis on the subjective measure (0.53 to 0.47), so that self-assessment might be more relevant in terms of judging an individual's risk level.

5.6.3 Robustness checks

The secondary results can be partly read as sensitivity analysis, but only rigorous robustness checks are presented here. We use OLS specifications since the Poisson specifications create conversion issues for a few of the following checks. We prefer at this point to have a homogeneous regression set-up that allows for comparative results.

Results by groups

We start by clustering the full sample size into groups. Table 5.12 on page 160 performs regressions with the same covariates as before with the full sample in regression (1) and the following subsamples: (2) farming households, (3) households that work in agriculture as well as day labor, (4) households that do not work in agriculture, (5) households from the lowest two income quintiles, (6) households from the lowest two wealth quintiles, (7) Hindu households, and (8) Muslim households. The results for the preference coefficients are mixed. For the poorest 40% of the households, measured either by income or by assets, the effect of risk preference is insignificant, and neither is altruism significant for the poorest 40% with regard to wealth. Risk preference has neither an effect for households that do *not* work in agriculture-related jobs (although the subsample is rather small with below 10% of the initial sample; hence, the results are not very robust themselves) nor for households that are Muslim.

On the other hand, for households that work in agriculture and for Hindu households, the effect of risk preference is highly significantly and can be replicated in regard to effect size. For farming households the effect size is even stronger. In regard to altruism, each subsample aside from households in non-agricultural jobs shows a positive effect, which is quite robust in terms of significance and effect size.

The relevance of the full models, represented by the Adjusted R^2 in the discussed Table 5.12 on page 160, indicates that the model with the included independent variables might not be sufficient for explaining dietary diversity for the poorest households in particular. In the next robustness check we therefore vary and add covariates.

Varying controls

In the Appendix, Tables D.2 - D.5 on pages 227 - 231 present four different regressions in which we changed either individual, household, village or all of these control variables. The sample sizes per regression might vary due to possible missing values for some variables. With any combination of alternative variables, risk preference and altruism continue to stay highly significant. The coefficients for risk vary from 0.081 to 0.204 and for altruism from 0.148 to 0.168.

Multicollinearity checks

As mentioned earlier, due to our estimation strategy, the regressions might be prone to multicollinearity issues, which potentially causes us to overestimate the effect sizes. The regression result from the main specification is checked for multicollinearity by computing the Variance Inflation Factor (VIF). The mean VIF of

Table 5.12: Correlation of various groups

	(1)	(2)	(3)	(4)	(5)	(6)	(7)	(8)
	OLS	OLS	OLS	OLS	OLS	OLS	OLS	OLS
Dependent variable	(log) DDW	(log) DDW	(log) DDW	(log) DDW	(log) DDW	(log) DDW	(log) DDW	(log) DDW
Risk level of spouse	0.085**	0.178***	0.096**	−0.000	0.015	−0.035	0.116**	−0.009
	(0.041)	(0.061)	(0.041)	(0.265)	(0.078)	(0.065)	(0.053)	(0.068)
Altruism of household head	0.162***	0.121	0.130**	0.389	0.096	0.199**	0.112	0.188*
	(0.061)	(0.085)	(0.062)	(0.299)	(0.110)	(0.098)	(0.076)	(0.104)
Individual covariates	yes	yes	yes	yes	yes	yes	yes	yes
Household covariates	yes	yes	yes	yes	yes	yes	yes	yes

Table 5.12: *(continued)*

	(1)	(2)	(3)	(4)	(5)	(6)	(7)	(8)
	OLS	OLS	OLS	OLS	OLS	OLS	OLS	OLS
Dependent variable	(log) DDW	(log) DDW	(log) DDW	(log) DDW	(log) DDW	(log) DDW	(log) DDW	(log) DDW
Village covariates	yes	yes	yes	yes	yes	yes	yes	yes
Observations	1102	609	1018	84	353	413	757	345
Adjusted R^2	0.239	0.302	0.249	0.224	0.139	0.131	0.316	0.077

Robust standard errors clustered by household in parentheses

* $p<0.10$, ** $p<0.05$, *** $p<0.01$

(1) includes full sample
(2) includes only farming households
(3) includes only households that work in agriculture
(4) includes only households that do not work in agriculture
(5) includes only households from the lowest 2 income quintiles
(6) includes only households from the lowest 2 asset quintiles
(7) includes only Hindu households
(8) includes only Muslim households

the regression is 1.85, which does not give rise to any concern in regard to multicollinearity issues. Table D.6 on page 233 of the Appendix presents the regression and the variable-specific VIF.

5.7 Conclusion

This chapter analyzes the effects that individual preferences have on nutrition intake. Hypotheses that are derived from an utility maximization model provide guidance for the research objective of this chapter, which is the empirical estimation of risk preference and altruism in regard to nutrition intake. The main results state that risk preference of the spouse has a positive, relevant and highly significant effect on her nutrition. The altruism of the household head has similarly a positive, relevant and highly significant effect on his spouse's nutrition. Other food intake indicators hold true to this relationship, so that risk preference and altruism in general affect the food and nutrition security of women and of the households they live in. The results are very robust to different controls and statistical checks. Hence, this chapter presents a novel viewpoint on food and nutrition security by introducing core concepts of behavioral economics to the debate on diminishing malnutrition.

Under uncertain conditions - in which human labor is the dominant source of individual income and where shocks can have potentially devastating effects due to missing institutions such as safety nets - risk takers will increase their nutrition marginally by 0.9% to 1.4% on average in the current period. The reasoning can be explained as follows: risk takers will forgo any safety considerations despite possible future negative shocks and consume more in the current period rather than preferring to save. In other words, they invest in their current nutrition in the hope of a better return in the future. This research does not answer the question of whether this behavior is intentional or subconscious. Similarly, in a setting where food is generally consumed communally within the household, the household head eats first, followed by his children and lastly by his wife. Altruistic behavior of the household head, which can be articulated as leaving more food for the rest of his family including his wife, increases the nutrition of women by 1.1% to 3.0%.

Whereas development economics often discuss approaches to either increase the access or the availability of resources for improving food and nutrition security, we discussed the question of how food is utilized more effectively in terms of nutritional intake on the basis of purely personal preferences. Risk preference is often elicited to understand investment decisions. Nutrition has to be regarded as such an investment decision, particularly in rural areas of marginalized communities, where a healthy physical status is often the primary requirement for income generation. Altruism is similarly a concept that has been used widely for understanding interpersonal behavior and effects of social cohesion, but rarely in terms of intrahousehold food sharing. In view of the very robust and relevant results, it is striking that the literature lacks theoretical and empirical work in this interdisciplinary endeavor. Utilizing the positive effects for reducing malnutrition and its effects seems called for.

Given the nature of individual preferences, policy implications are unfortunately not easily deducted. The present study uses cross-sectional data for the analysis reflecting the situation at one point in time. Even though deep preferences tend to be relatively stable, the environment and personal situation influence these. Hence, policies that are deduced from the current setting and that reflect an optimal strategy on the basis of the current setting, might potentially have suboptimal outcomes given their preference changing effects. Therefore, the implication is foremost that economic preferences do also have an effect on nutrition. The plans of food and nutrition security programs might want to take into account these effects for developing possible activities or for assessing the impact of these activities. Moreover, this study is a case in point in proving that revealing deep preferences is possible in a resource-effective way in terms of costs and time. Therefore, the results imply that targeting households with certain characteristics is possible for effective policy/program implementation. For instance, risk averse households might need more intensive support in nutrition, whereas risk taking households need a stronger support for possible coping measures.

We end this study with an outlook for further research in this realm. The empirical results show a stronger and even more robust effect of altruism on nutrition. Theoretical models on nutrition have not yet fully integrated altruism. Therefore, the effects that altruism can have need to be more highlighted and identified by theoretical models. Also, it is worthwhile to further explore empirical effects on various nutrition intake indicators, or possibly on nutrition outcome indicators - presuming a strong causal link. Altruism is a widely discussed concept in the economic literature. Given the glimpse provided by this study, further attempts at bridging the disciplines and examining the impact on food and nutrition security seem promising.

6 Concluding Remarks

This chapter concludes the presented research and serves three purposes. Firstly, the key research findings are highlighted and summarized. Secondly, implications for policy design are given. Thirdly, the limitations of the research are discussed in connection with implications for further research. Each chapter has specific conclusions; therefore, at this point, we draw the results together and aim for a broader discussion.

6.1 Summary and Contribution

Food and nutrition insecurity is prevalent across the globe and specifically in South Asia (FAO, IFAD, UNICEF, et al., 2018). India has high malnutrition rates despite having an economic surge over the past 25 years (DeLong, 2003). Rural areas are most concerning in regard to food and nutrition insecurity as well as in regard to lacking economic opportunities. In the rural areas, more than half of the population suffers from hidden hunger and around 40% of all children are food insecure. At the same time obesity rates are steadily increasing with currently 15% of the rural population being affected (ICF, 2015). Inadequate diets are often a direct cause of these malnutrition rates (Biesalski, 2013). Many factors force individuals to consume suboptimal diets, for instance unavailability or inaccessibility of foods. Food selection and consumption is also often a choice, even within a limited variety of food items. The present research identifies factors that influence consumption choices and estimates their effects on food and nutrition security. The focus is on diverse diets as a prerequisite for micronutrient adequacy.

Building upon the FAO and UNICEF frameworks for food and nutrition security (FAO, IFAD, and WFP, 2015; UNICEF, 1990), we have conceptualized links of relevant factors for food consumption. Households and individuals have an endogenous demand for food that is influenced for example by demographics, preferences or traditions. Food availability is mostly a combination of self-production of food or market availability. Embedded in an exogenous environment, households and individuals can realize their dietary demands to a certain extent dependent upon their accessibility to the available foods.

One descriptive and three analytical chapters have discussed four aspects of the conceptual framework: the economic environment, production choices, market accessibility and individual preferences. The case of rural India is taken for the empirical analysis. It is based on a household survey that was conducted in the states of Jharkhand, Karnataka and West Bengal.

Chapter 2 - *Focus on India: Food Prices and Food Security* - introduces the study environment in India and associates one major driver of food consumption choices with the food and nutrition security situation: prices (Bouis et al., 2011). Agricultural research has increased productivity of cereals since the 1970s, a process that is commonly referred to as the Green Revolution. The yields were able to nourish India's population growth and to reduce the risk of famines. But prices of other food groups such as vegetables and fruits increased relatively to cereals, setting a price incentive to consume less diverse diets (Government of India, 2018a). At the same time, total consumption of more expensive food groups has increased including both healthy (e.g. fruits and vegetables) and unhealthy items (e.g. sugar and vegetable oils). The higher consumption of these foods is linked with the income increase that India experienced since the 1990s; however, this is limited mostly to urban areas (World Bank, 2017). Accordingly, the malnutrition situation of today displays itself as the dual burden of malnutrition. Large parts of the rural population in particular suffer from undernutrition, whereas a growing number of the urban population suffers from overnutrition[1]. At the same time urban food consumption trends also tend to penetrate rural areas. Processed foods are often consumed in urban areas with a higher density of supermarkets or smaller convenient shops (e.g. Chege et al., 2015). Rural areas have a different retail structure, yet, low-cost processed items are increasingly consumed (Reardon et al., 2012). This adds to the picture that overnutrition in rural areas is appearing as well (ICF, 2015). Complicating the situation for the rural population are the few economic opportunities aside from the agricultural sector that could enable a higher income and, thus, the ability to afford the steadily increasing prices of healthier foods (Gupta et al., 2018).

Chapter 3 - *Production Diversity and Diets* - considers farming households and analyzes the causal links between production choices, consumption choices and the intermediary factor of market access (Jones, 2017; Sibhatu and Qaim, 2018b). Nutrition-sensitive agriculture is regarded as a decisive factor for reducing micronutrient malnutrition in that a diverse production enables a diverse consumption (Hoddinott, 2012; Pinstrup-Andersen, 2013). A conceptual framework and a theoretical framework are presented to illustrate the possible channels. By using multiple regression and the instrument approach with the variance

---

1   31.4% of all women in urban areas and 15.1% in rural areas are overweight, whereas 15.5% of all women in urban areas and 26.8% in rural areas are underweight (ICF, 2015)

of rainfall as an instrumental variable, the significance and relevance of the production-nutrition link is estimated. A positive association is found in that 1 additional food group produced relates up to a 18.8% increase in consumed food groups for women. Market access reduces the relevance and becomes the main driver of diverse diets when markets are within reach. However, primarily higher income groups benefit from the positive effect of markets, and the positive effect can only be related to the food groups of dairy products, vegetables and nuts. This chapter presents a contribution in theory and methodology by determining the consumption choices conditional on a certain set of production choices given the degree to which markets are within reach.

Chapter 4 - *Considering Preferences for Food Consumption* - takes a closer look at the individual decision making process and how consumption choices are formed based on individual preferences. A theoretical model is presented that amends the expected utility maximization framework with the economic preferences risk, patience and altruism. These preferences are influential for economic behavior and can explain seemingly irrational behavior, which can generally not be explained with the standard model of utility maximization (Kahneman, Knetsch, et al., 1990; Thaler and Sunstein, 2008). In the intertemporal model, nutrition is regarded as investment toward improving the ability of income generation in an uncertain future. Risk, patience and altruism shape individual budget allocation choices either towards nutrition or towards other goods. The model's optimal solution hypothesizes a positive impact of risk on current nutrition and a positive impact of altruism on another household member's nutrition.

Chapter 5 - *Effects of Preferences on Food Consumption* - tests the model's hypotheses in the setting of rural India. A survey tool is developed that elicits the preferences of the respondents with hypothetical games, enabling a link to the food consumption of the respondents and their households. Multiple regressions estimate the extent of the effects of risk and altruism on various nutrition indicators. It is estimated that an increase of 10 percentage points in risk-taking increases the dietary diversity score by 0.9% to 1.4%. Altruistic behavior of the household head improves the nutrition of his spouse by 1.1% to 3.0%. Chapters 4 and 5 have contributed to the literature in describing the theoretical link that economic preferences have on food and nutrition security, and by empirically confirming and quantifying this relationship.

6.2 Policy Implications, Limitations and Further Research

Policy implications aiming for an improvement of individual nutrition must consider the current situation: Micronutrient malnutrition is widespread and rural areas are lacking economic opportunities for large parts of the population. Market integration seems a key aspect for better dietary diversity, yet considering market access as a goal in itself falls short of considering the livelihoods of marginalized communities. The present research can give indications for possible policy channels, but limitations inherent to the research require a more intensive analysis before actual policy recommendations can be formulated. Therefore, the policy implications are given conditional upon various limitations that themselves bear implications for further research.

Policy Implications

Throughout the empirical results, income proved to be a highly significant and robust variable in forming a positive relation to nutrition. This is hardly surprising in the face of steadily rising food prices and general food access considerations. In rural areas, income is mostly generated through agriculture-related activities, which can be on-farm activities or take the form of day labor activities. 61% of all household heads recorded farming as their primary occupation and 18% considered themselves as agricultural workers. The agricultural produce is predominantly sold at local markets; hence, income is generated directly by selling agricultural products locally or indirectly by associated wages. There are opportunities to improve income through this channel, generally speaking, by reducing costs of production. We discussed the factor of transaction costs. These are relatively higher for high-value crops such as vegetables and fruits. Improved preservation techniques or reduced transportation costs can reduce the overall costs. Accordingly, these approaches can increase profits while keeping food prices at the same level. Respectively, market dynamics can lead to lower food prices while profits stay at a steady level.

When considering market integration as the share of agricultural products that is sold at markets, not all farmers are equally integrated into markets. The results indicate that production diversity has a stronger association with dietary diversity if market access is not a given. Therefore, improving the variety of food groups produced can be a channel for improvements for marginalized farming households. On the policy level, incentives for underutilized crops could be a rather macro-level mechanism, e.g. by subsidizing seeds or certain land use patterns, although micro-level programs that encourage a diversification of production are easier to implement. Nutrition-sensitive activities are frequently integrated by development programs and are often effective. Agricultural extension can bridge the policy and program levels. Unfortunately, 52% of the respondents reported only a satisfactory to very bad support by the governmental extension services. Hence, there is the potential to improve the quality of direct support by the extension service, at the same integrating nutrition-sensitive approaches.

The public support system is already quite extensive in the realm of food and nutrition security, partly induced by the famines in the first half of the 20th century. The results indicate the robust positive effects of the programs. Among the most frequently used programs are the Public Distribution System (PDS) (by 95% of all households), Anganwadi centers (65%) and the National Rural Employment Guarantee (NREG) (54%). Malfunctioning of these systems is anecdotally indicated and also formally reported (Government of India, 2017; Menon et al., 2017). Examples of the malfunction are inaccessibility to the food rations, too few employees at Anganwadi centers or no disbursements of guaranteed wages. As these schemes seem to have a vital function in securing food and nutrition security, any improvement in their efficiency is supportive.

Generally speaking, nutrition education is key to conscious food consumption. The presented preference model considers nutrition as investment with the assumption that the benefits of nutrition and the diminishing marginal returns are known. To some extent this assumption holds true, although detailed knowledge of the effects of micronutrients cannot be presupposed. Nutritional knowledge is mostly spread by Anganwadi workers and occasionally by activities of NGOs or other private organizations. Considering also that formal education proves to be a decisive factor for better income and, at the same time, for better nutrition, it seems natural to combine formal and nutrition education. There are already attempts to integrate nutrition education in school curricula (Government of India, 2017). A stronger integration with formal education can offer the opportunity to a better nutrition in that the likelihood for a higher income is increased as well.

Finally, the results indicate the positive effects of risk preference and altruism on dietary intake. Conversely, risk averse individuals or less-altruistic peers bear the likelihood of consuming a lower dietary quality. The research also shows that preferences can be elicited in large numbers despite resource limitations. Hence, by identifying risk averse and less altruistic individuals, a more focused support for nutrition can be provided. Accordingly, policy makers and local initiatives alike can utilize the methodology to improve the targeting of nutrition activities. Furthermore, didactic approaches for nutrition activities can also be tweaked to consider preferences for improving their effectiveness. For example, didactic games with varying payoffs might be more engaging for risk loving than for risk averse individuals.

Limitations and Further Research

The presented research results open up the opportunity for additional research questions and for improved results. Some concrete considerations are discussed below.

The data collection was built upon a cross-sectional study. This study design falls short of assessing temporal relationships between the variables of interest. The results can only be interpreted as a current state; time dynamics are not sufficiently reflected. We tried to overcome this challenge with recall information, but recalls can be incomplete. The instrumental variable approach is a technique to consider the omitted variable bias, but it has limitations in itself as discussed in Section 3.5.2. Longitudinal and panel data is advantageous in that unobserved heterogeneity can be controlled for. Regarding the relationship between production diversity and dietary diversity, publications using panel data are still missing (compare Sibhatu and Qaim, 2018b). Studies with cross-sectional data, however, confirm the presented results. Nevertheless, panel data is superior and should be used in further research, particularly when considering the effects of preferences on nutrition for which no comparable results exist in the literature.

Micronutrient malnutrition is widely assessed by using proxy indicators such as the dietary diversity scores. These are the second best solution as discussed in Section 1.5. A true assessment of micronutrient deficiencies that eliminates many shortcomings of the proxy indicators is the usage of biochemical indicators. These indicators entail concerns in regard to costs, technical implementation, ethics and individual consent, which is why we used proxy indicators as dependent variables. With sufficient time and resources, further research in the realm of food and nutrition security can be improved in precision and robustness by using biochemical assessments.

The focus of this research is set on the micro-level scale by analyzing individual and household behavior. The macroeconomic environment such as food prices are regarded as exogenous. However, the estimated effects of production diversity or altruism can have more far-reaching effects than solely on the household level. Food prices of micronutrient-rich foods have increased; a higher supply of diverse foods can equally have a decreasing effect on food price levels. Hence, production diversity might have a much bigger effect on dietary diversity than estimated when considering a larger scale. Similarly, the literature discusses positive effects of altruism on informal insurance arrangements (Foster and Rosenzweig, 2001), which can smooth the effects of shocks on consumption. Therefore, additional research that considers groups, villages or regions can better estimate the effects of production choices and preferences on individual nutrition.

Market access has been discussed in a binary form, either having access or not. This proves useful and demonstrates clearly the separability condition. However, markets are quite diverse in various dimensions. Markets can be differentiated into informal and formal markets, in size and geographical outreach or accessibility. Considering the sizable effect of markets on food and nutrition security, it is advisable to analyze the different effects of market characteristics more rigorously.

As the discussion of links between preferences and food consumption is rather novel and explorative, we see a few points for improving the research and for better analyzing the effects. First, the theoretical model considers individual consumption choices, although individual choices are often influenced by household considerations. A household decision making model could help in further generalizing the theoretical framework. Further research can include, for example, the unitary household model that links the members by altruism (Becker, 1974) or bargaining household models (Lundberg and Pollak, 1993).

Second, previous literature has discussed the effects of risk preference on unhealthy lifestyles, consumption of alcohol and on consumption of unsafe foods (Anderson and Mellor, 2008; Barsky et al., 1997; Lobb et al., 2007). We have established the association to indicators of food and nutrition security. A closer inspection of the link to the consumption of specific food groups or items seems natural. Analyses focused on food items can reveal consumption trends and possible risks, for instance, in regard to food safety concerns.

Third, food and nutrition security relates most of the time to vulnerable groups such as children and women. We focused our analysis on the effects of women's nutrition. Children are a special case in that food consumption is up to a certain age completely dependent on other individuals. Disentangling the effect of the parents' or other household members' preferences on the children's diets seems useful especially in terms of intergenerational transfers and future pay-offs. The inclusion of the life cycle skill formation approach could help to model the value of nutrition as a human capital investment over a lifetime instead of only two periods (Heckman, 2006).

Fourth, we considered isolated effects of preferences to prove and estimate the case in point. However, linkages to other consumption decisions can be regarded as well. For instance, risk preference is shown to affect innovative behavior and other investment decisions (Chetty and Szeidl, 2007; Gatzweiler and von Braun, 2016). Therefore it seems worthwhile to investigate the perceived and realized trade-offs between expenditures for nutrition, expenditures for other investments, or expenditures for risk-mitigating systems such as insurances and savings groups.

Appendix A Supplementary Materials to Chapter 1

A.1 Prevalence of Micronutrient Deficiency

Figure A.1: Prevalence of anemia among women (total)

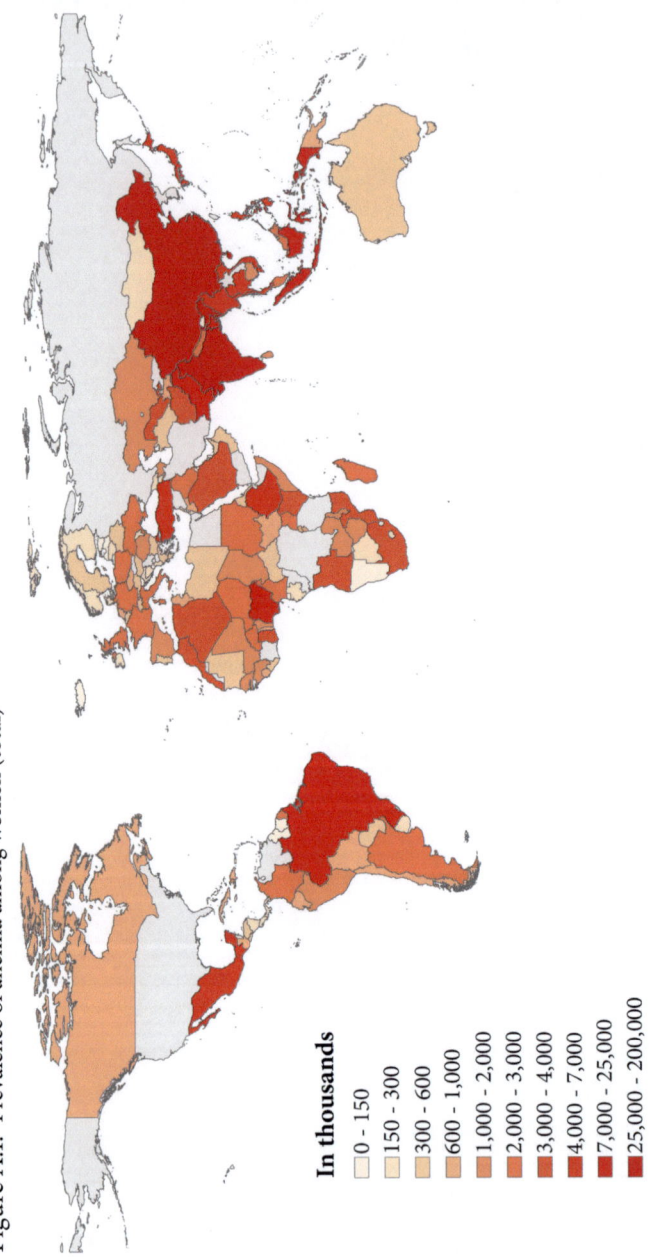

Data source: The World Bank (2017) anemic data for 2016, United Nations (2017) population data for 2015. Author's illustration

A.2 Micronutrients, Sources and Consequences of Their Lack in Nutrition

Micronutrient	Source	Consequences
Minerals		
Calcium	Eggs, bones	Premature delivery
Copper	Meat	Impaired physical development
Iodine	Fish, sea food	Cretinism, native deafness
Iron	Meat, pulses	Premature delivery, increased likelihood of maternal death
Magnesium	Meat	Premature delivery
Potassium, Sodium	Meat, eggs	Impaired physical development, premature delivery
Zinc	Meat, pulses	Impaired physical development, malfunctioning immune system
Vitamins		
Folic acid	Vegetables, meat	Neural tube defect, coloboma
Vitamin A	Meat, eggs	Malfunctioning immune system, birth defects
Vitamin B_1	Meat, sprouts	Impaired cognitive and physical development
Vitamin B_2	Meat, pulses, eggs	Impaired physical development
Vitamin C	Fruits, vegetables	Impaired connective tissue, malfunctioning immune system
Vitamin D	Synthesis in skin, some fish	Impaired physical development
Vitamin E	Fruits, sprouts	Fetal anemia

Sources: Biesalski, 2015; Wu et al., 2012

A.3 Food Groups of Food Intake Indicators

MDDW

1. Grains, white roots and tubers, and plantains
2. Pulses (beans, peas and lentils)
3. Nuts and seeds
4. Dairy products
5. Meat, poultry, fish
6. Eggs
7. Dark green leafy vegetables
8. Other vitamin A-rich fruits and vegetables
9. Other vegetables
10. Other fruits

WDDS

1. Starchy staples
2. Legumes, nuts and seeds
3. Dairy products
4. Organ meat
5. Meat and fish
6. Eggs
7. Dark green leafy vegetables
8. Other vitamin A rich fruits and vegetables
9. Other fruits and vegetables

MDD

1. Grains, roots and tubers
2. Legumes and nuts
3. Dairy products
4. Meat, poultry, fish
5. Eggs
6. Vitamin A-rich fruits and vegetables
7. Other fruits and vegetables

HDDS

1. Cereals
2. Root and tubers
3. Vegetables
4. Fruits
5. Meat, poultry, offal
6. Eggs
7. Fish and seafood
8. Pulses/legumes/nuts
9. Dairy products
10. Oil/fats
11. Sugar/honey
12. Miscellaneous

A.4 FIES Questions

1. You were worried you would run out of food because of a lack of money or other resources?
2. You were unable to eat healthy and nutritious food because of a lack of money or other resources?
3. You ate only a few kinds of foods because of a lack of money or other resources?
4. You had to skip a meal because there was not enough money or other resources to get food?
5. You ate less than you thought you should because of a lack of money or other resources?
6. Your household ran out of food because of a lack of money or other resources?
7. You were hungry but did not eat because there was not enough money or other resources for food?
8. You went without eating for a whole day because of a lack of money or other resources?

A.5 Random Walk Sampling Technique

The following lists stepwise the procedure for the random walk sampling in villages of West Bengal and Karnataka.

1. Identify the location of the village leader. Go to the village leader's location to introduce yourself and the study. Mention what you will do and how and that you would like to have a Focus Group Discussion with him and other village inhabitants later on. Then start your random walk.
2. Go from the village leader's location into the direction of the village center.
3. Chose the third household (left + right) on the way to ask if the sampling criteria (the household has at least one child below 2 years of age) is met in this household. If the sampling criteria is met, ask if you could do the survey with them by explaining how much time it will take and what you will do. If they agree, start the survey.
4. If the sampling criteria is not met, repeat Step 3.
5. If the sampling criteria is met, but the household does not want to respond right now to the questionnaire, ask if you can come back at another time when it is more suitable for the household. Note down the mobile number and come back at the agreed time to do the survey.
Then repeat Step 3.
6. If the sampling criteria is met, but the household does not want to respond at all. Ask for the reason why and note down a few key words that indicate the household's characteristic (solely by observation). E.g. how is the house built, how many people are in the household, is it a certain category (ST, SC), do the respondents look healthy etc.
Then repeat Step 3.
7. Once you come to an intersection, throw a coin to determine in which direction you go. Number means to the right, and symbol means to the left.
8. You continue this random walk in the village until you have reached the sample size of the village as indicated on the sampling document!
9. If you come to the end of the village before you reach the sample size, go back to the center of the village, and walk in a different direction than before.
10. If there are not sufficient households in the village that meet the sampling criteria and that want to participate in the survey, report to the Supervisor and explain the situation. The Supervisor will give further instructions.

A.6 Sampling

Table A.1: Sampling in India

	Jharkhand	West Bengal	Karnataka
Households	490	402	432
Individuals	2926	1750	1931
Children below 2	429	407	20
Children below 5	748	495	78
Villages	49	35	35
Districts	3	4	6
Markets	8	5	8

Figure A.2: Surveyed households in the regions of Jharkhand and West Bengal (borders represent districts)

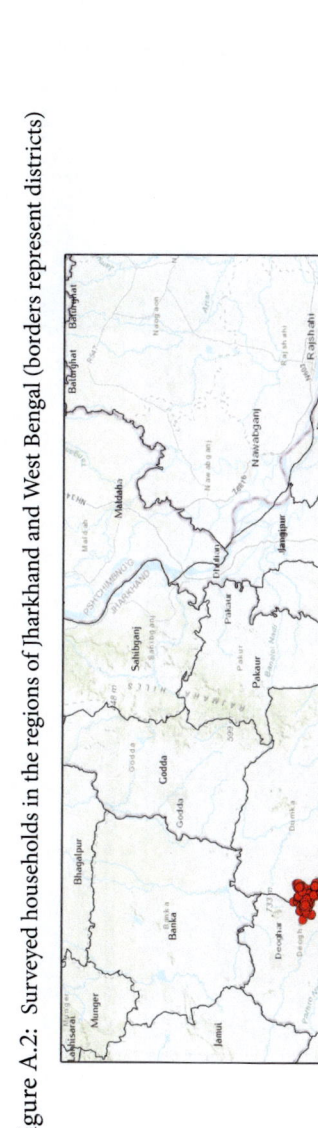

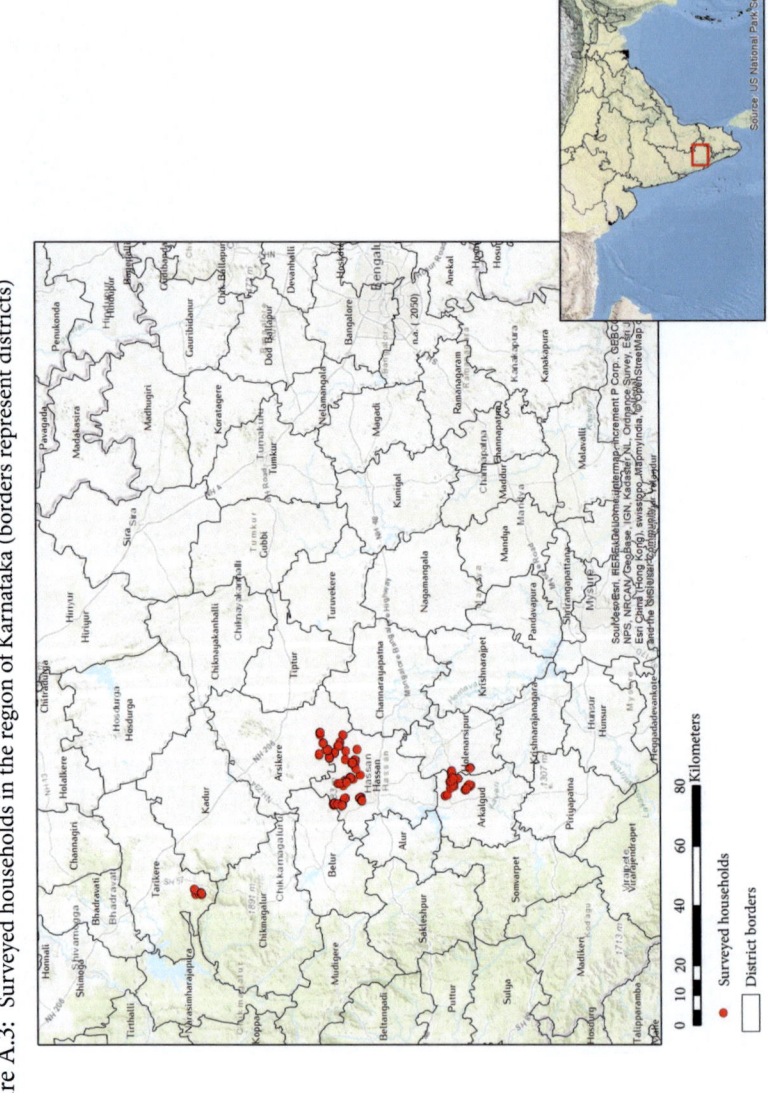

Figure A.3: Surveyed households in the region of Karnataka (borders represent districts)

Appendix B Supplementary Materials to Chapter 3

B.1 Checks for Instrument Variable

Table B.1: First stage regression results based on the linear model

Dependent variable: Production diversity	
Excluded instrument:	
Variance of rainfall (mm)	0.00002***
	(0.000)
Included instruments[y]**:**	Yes
R-squared	0.114
Weak-identification test	
Sanderson-Windmeijer F-statistic	13.66
Observations	761

Robust standard errors clustered by household in parentheses
* $p<0.10$, ** $p<0.05$, *** $p<0.01$
[y] Coefficients omitted to preserve space

Table B.2: Test for significant effects between rainfall variation and covariates

Dependent variable	(1) OLS (log) DDW	(2) OLS (log) Income	(3) OLS Formal education	(4) OLS Nonfarm occupation	(5) OLS Government schemes
Variance of rainfall	0.000 (0.000)	0.000 (0.000)	0.000 (0.000)	−0.000 (0.000)	0.000 (0.000)
Individual covariates:	Yes	Yes	Yes	Yes	Yes
Household covariates:	Yes	Yes	Yes	Yes	Yes
Village covariates:	Yes	Yes	Yes	Yes	Yes
Region fixed effects:	Yes	Yes	Yes	Yes	Yes
Observations	761	761	761	761	761
Adjusted R^2	0.296	0.488	0.476	0.301	0.248

Robust standard errors clustered by household in parentheses
*** $p<0.01$, ** $p<0.05$, * $p<0.1$

B.2 Effects of Policies

Table B.3: Effects of policy schemes usage on dietary diversity

	(1) OLS	(2) Poisson
Dependent variable	(log) DDW	DDW
PDS (1 = yes)	0.008	0.007
	(0.058)	(0.049)
BPL Card (1 = yes)	−0.020	−0.015
	(0.034)	(0.030)
Antodaya (1 = yes)	0.115	0.093
	(0.074)	(0.065)
NREGS (1 = yes)	0.038**	0.046**
	(0.026)	(0.023)
Indira Awas Yojana (1 = yes)	−0.012	0.013
	(0.053)	(0.042)
Mid-day meals (1 = yes)	0.063	0.044*
	(0.030)	(0.024)
Anganwadi (1 = yes)	−0.044	−0.054
	(0.042)	(0.033)
Widow pension (1 = yes)	0.031	0.023
	(0.049)	(0.042)
Old-age pension (1 = yes)	0.058	0.039
	(0.040)	(0.031)
Individual covariates:	Yes	Yes
Household covariates:	Yes	Yes
Village covariates:	Yes	Yes
Region fixed effects:	Yes	Yes
Observations	810	810
Adjusted R^2	0.291	
Pearson goodness-of-fit		487.84
p-value		1.00

Robust standard errors clustered by household in parentheses
*** $p<0.01$, ** $p<0.05$, * $p<0.1$

B.3 Market Access Effects

Figure B.1: Predictive margins of market visits with confidence intervals at 95%

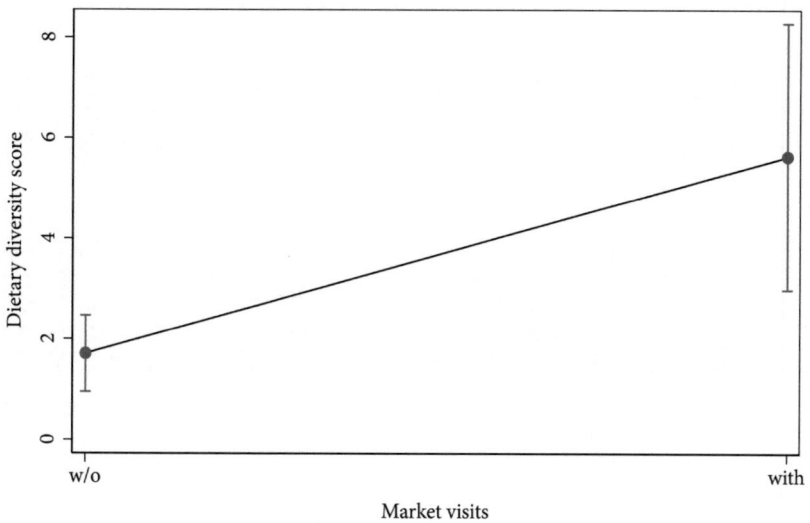

Table B.4: Impact of market access on dietary diversity of women with detailed results

	(1) OLS	(2) Poisson	(3) Linear IV (2SLS)	(4) Poisson IV (GMM)
Dependent variable	(log) DDW	DDW	(log) DDW	DDW
PD (A)	0.047***	0.045***	0.413**	0.386***
	(0.017)	(0.016)	(0.186)	(0.114)
PD X Market visit (B)	−0.045**	−0.042**	−0.438**	−0.361**
	(0.020)	(0.019)	(0.219)	(0.141)
Market visit (C)	0.147**	0.159***	1.323**	
	(0.066)	(0.062)	(0.654)	
Individual variables				
Age of woman	−0.009	0.020	−0.131	−0.121
	(0.065)	(0.058)	(0.102)	(0.076)
Literacy of woman	0.081***	0.057**	0.023	0.016
	(0.030)	(0.027)	(0.048)	(0.035)
Household variables				
Age of household head	0.003**	0.002	0.002	0.002
	(0.001)	(0.001)	(0.002)	(0.001)
Formal education of household head	0.001	0.001	0.006	0.005
	(0.004)	(0.003)	(0.005)	(0.004)
(log) Income	0.099***	0.098***	0.103***	0.069***
	(0.022)	(0.019)	(0.028)	(0.023)

Table B.4: *(continued)*

Dependent variable	(1) OLS (log) DDW	(2) Poisson DDW	(3) Linear IV (2SLS) (log) DDW	(4) Poisson IV (GMM) DDW
Religion (0 = Hindu, 1 = Muslim)	-0.021 (0.058)	-0.036 (0.056)	-0.078 (0.103)	-0.137 (0.117)
Number of males 0-5 years	0.008 (0.024)	0.009 (0.022)	0.006 (0.033)	0.011 (0.028)
Number of males 5-15 years	0.023 (0.019)	0.015 (0.017)	0.020 (0.026)	0.003 (0.024)
Number of males 15-60 years	-0.019 (0.018)	-0.015 (0.016)	-0.030 (0.022)	-0.020 (0.017)
Number of males 60+ years	0.004 (0.037)	-0.003 (0.031)	0.061 (0.054)	0.045 (0.041)
Number of females 0-5 years	-0.008 (0.025)	0.005 (0.022)	-0.031 (0.033)	-0.043 (0.034)
Number of females 5-15 years	-0.006 (0.019)	-0.003 (0.016)	0.010 (0.029)	0.004 (0.023)
Number of females 15-60 years	0.031* (0.019)	0.025 (0.016)	0.039* (0.021)	0.029* (0.017)
Number of females 60+ years	-0.028 (0.041)	-0.009 (0.034)	-0.006 (0.049)	0.012 (0.037)

Table B.4: *(continued)*

Dependent variable	(1) OLS (log) DDW	(2) Poisson DDW	(3) Linear IV (2SLS) (log) DDW	(4) Poisson IV (GMM) DDW
Nonfarm occupation	−0.011	0.010	−0.039	−0.055
	(0.043)	(0.040)	(0.054)	(0.051)
(log) Total land size	0.030*	0.025*	0.013	−0.008
	(0.016)	(0.015)	(0.030)	(0.027)
Government schemes	0.023**	0.025***	0.029**	0.025**
	(0.009)	(0.009)	(0.013)	(0.012)
Village variables				
(log) Village population	−0.006	−0.025	0.040	0.038
	(0.020)	(0.017)	(0.033)	(0.026)
Years that village is electrified	0.002	0.002	0.003	0.001
	(0.002)	(0.001)	(0.002)	(0.001)

Table B.4: *(continued)*

Dependent variable	(1) OLS (log) DDW	(2) Poisson DDW	(3) Linear IV (2SLS) (log) DDW	(4) Poisson IV (GMM) DDW
Regions				
West Bengal	0.329***	0.392***	0.391***	0.405***
	(0.076)	(0.072)	(0.145)	(0.136)
Karnataka	0.219***	0.256***	0.210**	0.278***
	(0.078)	(0.072)	(0.104)	(0.085)
Observations	807	807	756	756
Adjusted R^2	0.288			
Pearson goodness-of-fit		287.98		
p-value		1.00		
Sanderson-Windmeijer F-test (A)			17.98	
p-value			0.0000	
Sanderson-Windmeijer F-test (B)			10.66	
p-value			0.001	

Robust standard errors clustered by household in parentheses
*** $p<0.01$, ** $p<0.05$, * $p<0.1$

Table B.5: Correlations between market access and food group consumption

	(1) Probit Mfx / SE	(2) Probit Mfx / SE	(3) Probit Mfx / SE	(4) Probit Mfx / SE	(5) Probit Mfx / SE	(6) Probit Mfx / SE	(7) Probit Mfx / SE	(8) Probit Mfx / SE	(9) Probit Mfx / SE
Dependent variable	Legumes	Meat	Dairy	Nuts	Eggs	Leafy	Vitamin A-rich	Other vegetables	Other fruits
Market visits	−0.034 (0.118)	−0.005 (0.166)	0.098** (0.143)	0.013 (0.182)	0.003 (0.260)	−0.008 (0.127)	−0.015 (0.186)	0.087*** (0.120)	−0.007 (0.161)
Individual covariates:	Yes	Yes	Yes	Yes	Yes	Yes	Yes	Yes	Yes
Household covariates:	Yes	Yes	Yes	Yes	Yes	Yes	Yes	Yes	Yes
Village covariates:	Yes	Yes	Yes	Yes	Yes	Yes	Yes	Yes	Yes
Region fixed effects:	Yes	Yes	Yes	Yes	Yes	Yes	Yes	Yes	Yes
Number of observations	807	807	807	807	807	807	807	807	807
Pseudo R^2	0.066	0.154	0.265	0.265	0.398	0.319	0.200	0.032	0.199
LR chi2	60.950	65.018	186.391	194.884	94.675	281.937	103.547	23.370	103.810
Prob > chi2	0.000	0.000	0.000	0.000	0.000	0.000	0.000	0.498	0.000
Baseline predicted probability	0.699	0.067	0.292	0.130	0.029	0.616	0.082	0.807	0.132

Robust standard errors clustered by household in parentheses
*** p<0.01, ** p<0.05, * p<0.1

B.4 Production Diversity and Income

Figure B.2: Production diversity and income quintiles heat map

Production Diversity \ Income Quintile	1	2	3	4	5
1	2.22	2.34	3.7	3.08	2.84
2	0.37	0.25	0.49	0.99	0.74
3	9.12	10.97	9.86	11.84	18.25
4	1.97	1.85	2.1	3.21	5.55
5	0.86	1.23	0.74	0.62	1.36
6	0.74	0.74	0.37	0.12	0.12
7	0.74	0.12	0.12	0	0
8	0.25	0	0	0.12	0

Note: Percentages are given. Yellow reflects a low frequency, red reflects a high frequency.

Appendix C Supplementary Materials to Chapter 4

C.1 Solving Integral for Mean Variance Form

We assume the form of the utility function to be exponential, we also consider the measure of risk preference as A and a two-period model such that:

$$V_{c_0,c_1} = -e^{\{-Ac_0\}} - e^{\{-Ac_1\}}$$

Considering the utility equation 4.6:

$$V = -exp\{-A(y_0 - np - s_0)\}$$
$$- \int_{-\infty}^{\infty} \delta exp\{-A[w\mu f(n) + (1+r)(y_0 - np - c_0)]\}\varphi(x)dx$$

We substitute s_0 and use the chain rule for constants in the integral:

$$= -exp\{-A(y_0 - np - s_0)\}$$
$$-\delta \int_{-\infty}^{\infty} exp\{-A[w\mu f(n) + s_0(1+r)]\}\varphi(x)dx$$

For the probability density function $\varphi(x)dx$ of the shocks x, we assume a normal distribution with variance σ^2 and mean μ:

$$= -exp\{-A(y_0 - np - s_0)\}$$
$$-\delta \int_{-\infty}^{\infty} exp\{-A[w\mu f(n) + s_0(1+r)]\} * exp\left\{\frac{-(x-\mu)^2}{2\sigma^2}\right\}dx$$

Simplifying the equation:

$$= -exp\{-A(y_0 - np - s_0)\}$$
$$-\delta \int_{-\infty}^{\infty} exp\left\{-A[w\mu f(n) + s_0(1+r)] - \frac{-(x-\mu)^2}{2\sigma^2}\right\}dx$$

Solving the integral, we get:

$$= -exp\{-A(y_o - np - s_o)\}$$

$$-\frac{\sqrt{2}\sqrt{\pi}i}{2\sqrt{-\frac{1}{\sigma^2}}} \text{erf}\left(\frac{\sqrt{2}\left(Af(n)w\sigma^2 i - \mu i + x_1 i\right)}{2\sigma^2\sqrt{-\frac{1}{\sigma^2}}}\right)$$

$$* \delta \, exp\left\{-A\left(f(n)\mu w + (1+r)s_o - \frac{1}{2}Af(n)^2 \sigma^2 w^2\right)\right\}$$

By definition the error function to the infinity is 1, therefore we consider:

$$-\frac{\sqrt{2}\sqrt{\pi}i}{2\sqrt{-\frac{1}{\sigma^2}}} \text{erf}\left(\frac{\sqrt{2}\left(Af(n)w\sigma^2 i - \mu i + xi\right)}{2\sigma^2\sqrt{-\frac{1}{\sigma^2}}}\right) \approx 1$$

Hence, the final equation is:

$$V = -exp\{-A(y_o - np - s_o)\}$$

$$-\delta \, exp\left\{-A\left(f(n)\mu w + (1+r)s_o - \frac{1}{2}Af(n)^2 \sigma^2 w^2\right)\right\}$$

C.2 Differentiation for Preference Model

In the following, we will create the FOC of the expected utility over nutrition n and savings s. The expected utility is given in equation 4.7:

$$V = -exp\{-A(y-np-s)\}$$
$$-\delta exp\left\{-A\left(f(n)\mu w + (1+r)s - \frac{1}{2}Af(n)^2 \sigma^2 w^2\right)\right\},$$

FOC over n

Using the chain rule for the derivative of the composite equation:

$$\frac{dV}{dn} = -\left[\frac{d}{dn}\{-A(y-np-s)\}\right]exp\{-A(y-np-s)\}$$
$$-\left[\frac{d}{dn}\left\{-A\left(f(n)\mu w + (1+r)s - \frac{1}{2}Af(n)^2 \sigma^2 w^2\right)\right\}\right]$$
$$* \delta exp\left\{-A\left(f(n)\mu w + (1+r)s - \frac{1}{2}Af(n)^2 \sigma^2 w^2\right)\right\}$$

Differentiate the sum terms by term:

$$= -\left[\frac{d}{dn}\{-Ay + Anp + As\}\right]exp\{-A(y-np-s)\}$$
$$-\left[\frac{d}{dn}\left\{-Af(n)\mu w + A(1+r)s + \frac{1}{2}A^2 f(n)^2 \sigma^2 w^2\right\}\right]$$
$$* \delta exp\left\{-A\left(f(n)\mu w + (1+r)s - \frac{1}{2}Af(n)^2 \sigma^2 w^2\right)\right\}$$

Factor out constants and zeros:

$$= -\left[Ap\frac{d}{dn}n\right]exp\{-A(y-np-s)\}$$
$$-\left[-A\mu w\frac{d}{dn}f(n) + \frac{1}{2}A^2 \sigma^2 w^2 \frac{d}{dn}f(n)^2\right]$$
$$* \delta exp\left\{-A\left(f(n)\mu w + (1+r)s - \frac{1}{2}Af(n)^2 \sigma^2 w^2\right)\right\}$$

Taking the derivatives:

$$= -\left[Ap_1\right]exp\{-A(y-np-s)\}$$

$$-\left[-A\mu wf'(n) + \frac{1}{2}A^2\sigma^2 w^2 2f(n)f'(n)\right]$$

$$*\delta exp\left\{-A\left(f(n)\mu w + (1+r)s - \frac{1}{2}Af(n)^2\sigma^2 w^2\right)\right\}$$

Simplifying gets the final expression:

$$= -\left[Ap\right]exp\{-A(y-np-s)\}$$

$$+\left[A\mu wf'(n) - A^2\sigma^2 w^2 f(n)f'(n)\right]$$

$$*\delta exp\left\{-A\left(f(n)\mu w + (1+r)s - \frac{1}{2}Af(n)^2\sigma^2 w^2\right)\right\}$$

FOC over s

Using the chain rule for the derivative of the composite equation.

$$\frac{dV}{ds} = -\left[\frac{d}{ds}\{-A(y-np-s)\}\right]\exp\{-A(y-np-s)\}$$

$$-\left[\frac{d}{ds}\left\{-A\left(f(n)\mu w + (1+r)s - \frac{1}{2}Af(n)^2\sigma^2 w^2\right)\right\}\right]$$

$$*\,\delta\exp\left\{-A\left(f(n)\mu w + (1+r)s - \frac{1}{2}Af(n)^2\sigma^2 w^2\right)\right\}$$

Differentiate the sum terms by term:

$$= -\left[\frac{d}{ds}\{-Ay + Anp + As\}\right]\exp\{-A(y-np-s)\}$$

$$-\left[\frac{d}{ds}\left\{-Af(n)\mu w - As - Ars + \frac{1}{2}A^2 f(n)^2 \sigma^2 w^2\right\}\right]$$

$$*\,\delta\exp\left\{-A\left(f(n)\mu w + (1+r)s - \frac{1}{2}Af(n)^2\sigma^2 w^2\right)\right\}$$

Factor out constants and zeros:

$$= -\left[A\frac{d}{ds}s\right]\exp\{-A(y-np-s)\}$$

$$-\left[-A\frac{d}{ds}s - Ar\frac{d}{ds}s\right]$$

$$*\,\delta\exp\left\{-A\left(f(n)\mu w + (1+r)s - \frac{1}{2}Af(n)^2\sigma^2 w^2\right)\right\}$$

The derivative of s_0 is 1:

$$= -\left[A1\right]\exp\{-A(y-np-s)\}$$

$$-\left[-A1 - Ar1\right]$$

$$*\,\delta\exp\left\{-A\left(f(n)\mu w + (1+r)s - \frac{1}{2}Af(n)^2\sigma^2 w^2\right)\right\}$$

Simplifying gets the final expression:

$$= -\left[A\right]exp\{-A(y-np-s)\}$$

$$+\left[A(1+r)\right]$$

$$* \delta\, exp\left\{-A\left(f(n)\mu w + (1+r)s - \frac{1}{2}Af(n)^2 \sigma^2 w^2\right)\right\}$$

C.3 Solving for Optimal Savings Rate

We use the FOC $\frac{d}{ds}$ and find its vertex point for s (hypothesized maximum) by setting it equal to zero:

$$\frac{d}{ds} = -\left[A\right]\exp\{-A(y-np-s)\}$$

$$+\left[A(1+r)\right]$$

$$*\,\delta\exp\left\{-A\left(f(n)\mu w + (1+r)s - \frac{1}{2}Af(n)^2\sigma^2 w^2\right)\right\}$$

$$= 0$$

By taking the logarithm of the exponential and by changing the order of the equation, we get:

$$s = -\frac{\ln\left(\frac{1}{\delta+\delta r}\right) - A(y-np) - \frac{A^2 f(n)^2 w^2 \sigma^2}{2} + Af(n)\mu w}{2A+Ar}$$

We simplify the expression by approximating the term $\ln\left(\frac{1}{\delta+\delta r}\right)$ to zero, so we find the optimal savings rate at:

$$s^* = -\frac{-A(y-np) - \frac{A^2 f(n)^2 w^2 \sigma^2}{2} + Af(n)\mu w}{2A+Ar}$$

Appendix D Supplementary Materials to Chapter 5

D.1 Spatial Distribution of Preferences in the World

Figure D.1: Distribution of risk in the world

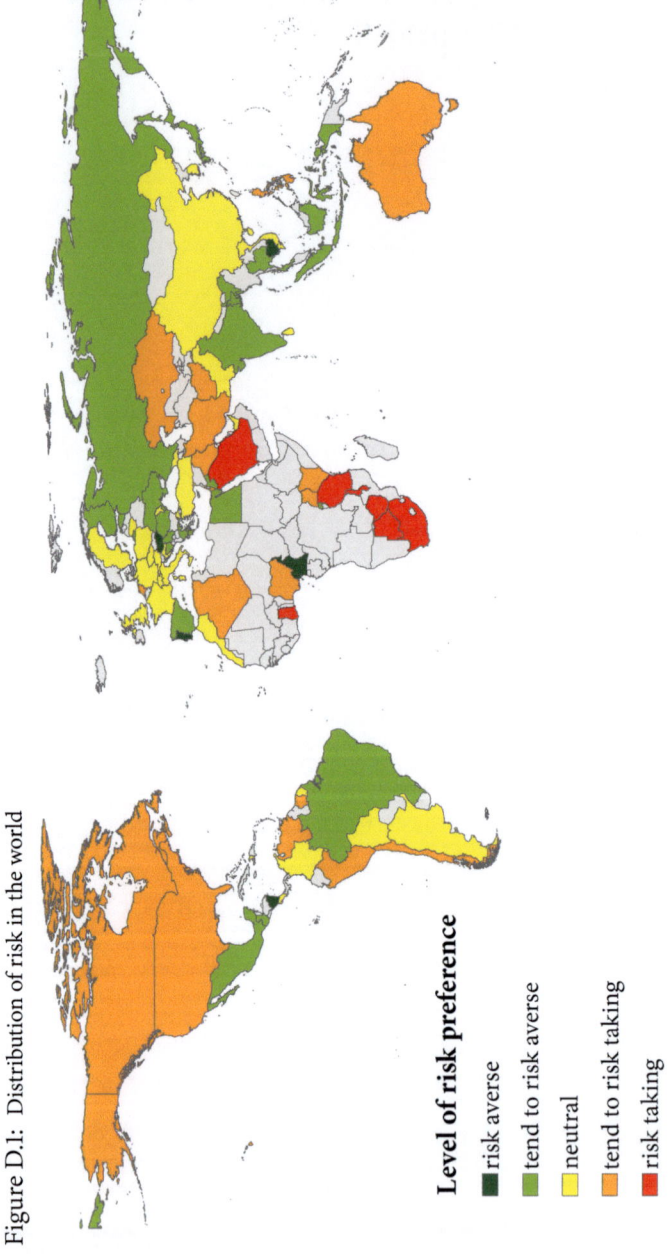

Data source: Falk et al., 2018; Falk, Becker, Dohmen, Huffman, et al., 2016. Author's illustration

Spatial Distribution of Preferences in the World 203

Figure D.2: Distribution of altruism in the world

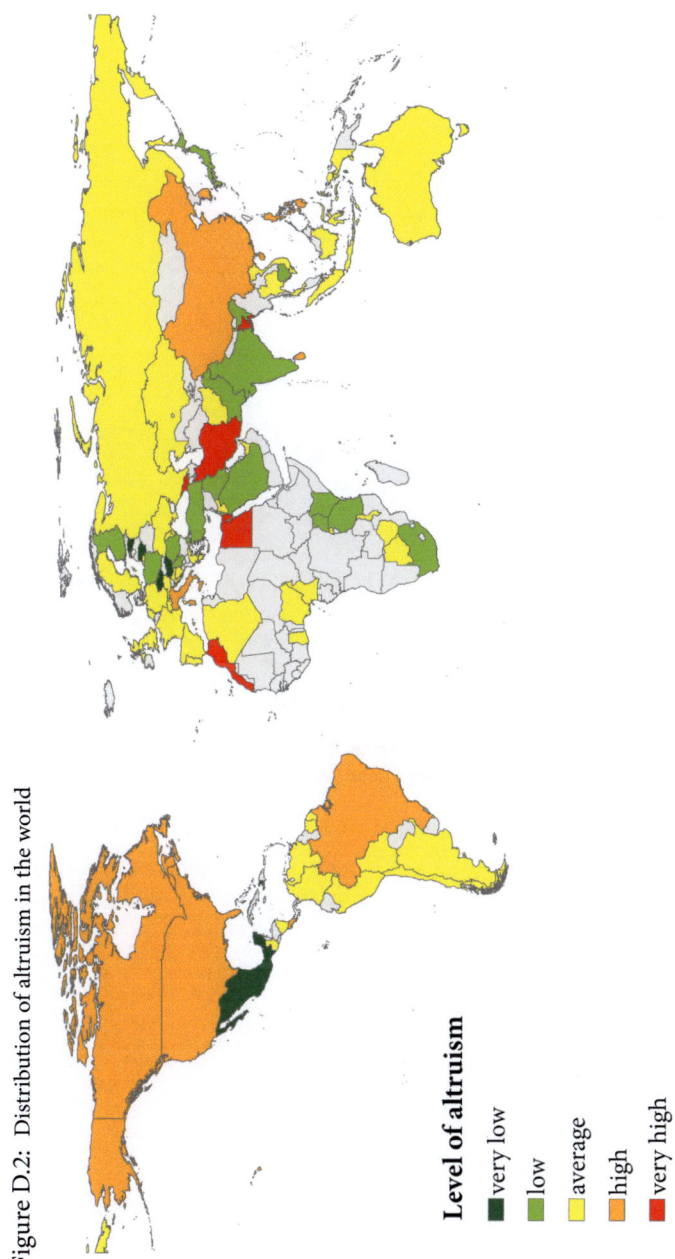

Data source: Falk et al., 2018; Falk, Becker, Dohmen, Huffman, et al., 2016. Author's illustration

D.2 Distribution of Preferences in the World

Figure D.3: Probability distribution of patience in the world

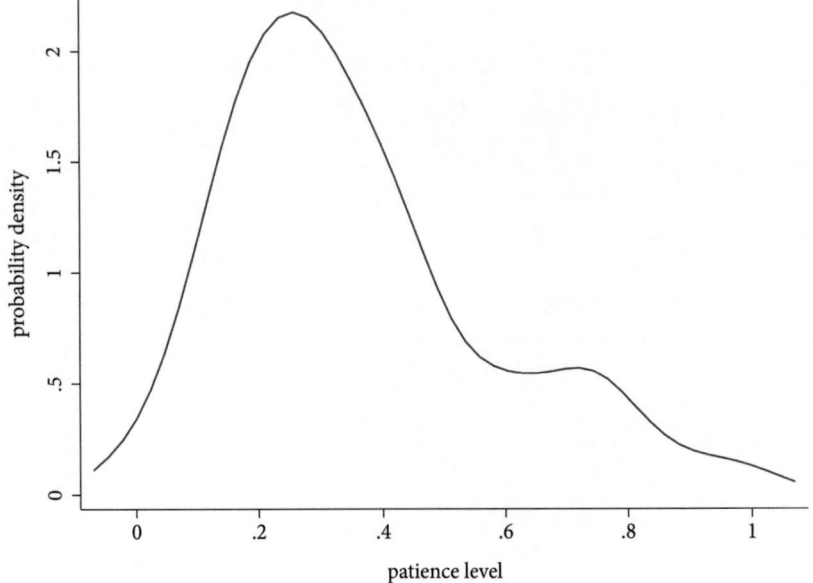

Data source: Falk et al., 2018; Falk, Becker, Dohmen, Huffman, et al., 2016. The distribution is computed as kernel density estimation using a Gaussian kernel

Figure D.4: Probability distribution of risk in the world

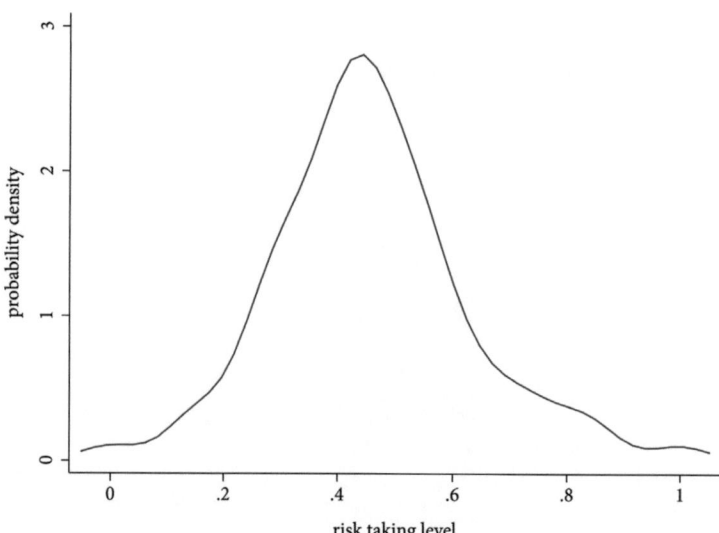

Data source: Falk et al., 2018; Falk, Becker, Dohmen, Huffman, et al., 2016. The distribution is computed as kernel density estimation using a Gaussian kernel

Figure D.5: Probability distribution of altruism in the world

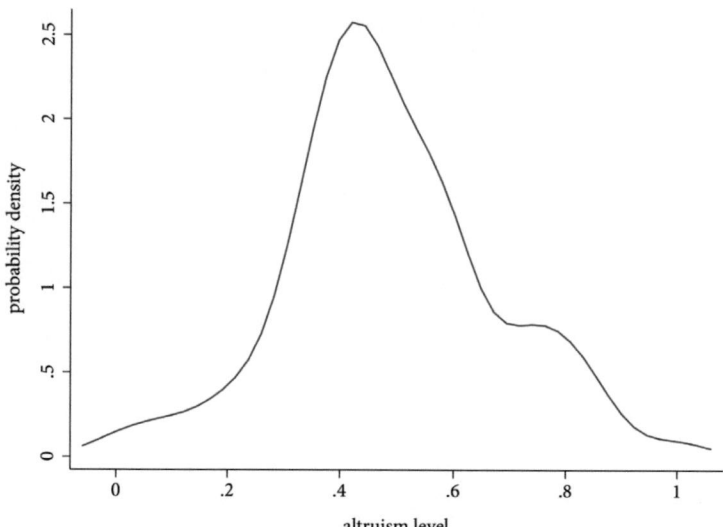

Data source: Falk et al., 2018; Falk, Becker, Dohmen, Huffman, et al., 2016. The distribution is computed as kernel density estimation using a Gaussian kernel

D.3 Preferences and Income in the World

Figure D.6: Correlation of risk and income globally

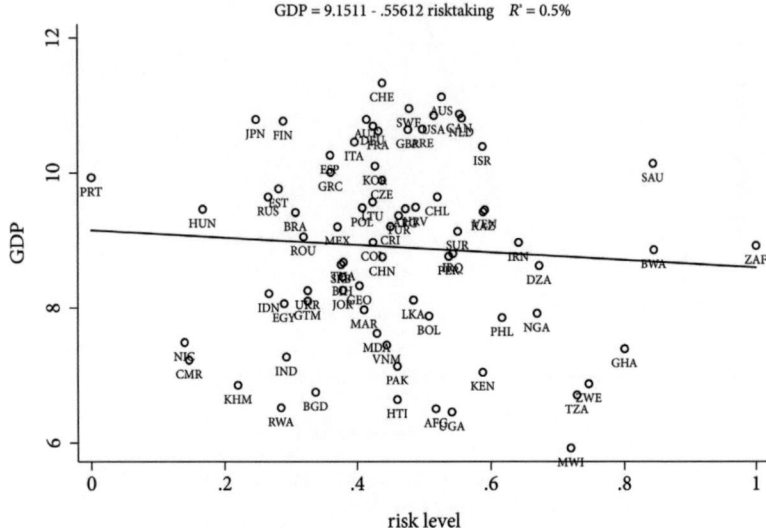

Note: log of GDP per capita in USD

Data source: Falk et al., 2018; Falk, Becker, Dohmen, Huffman, et al., 2016; World Bank, 2017

Figure D.7: Correlation of altruism and income globally

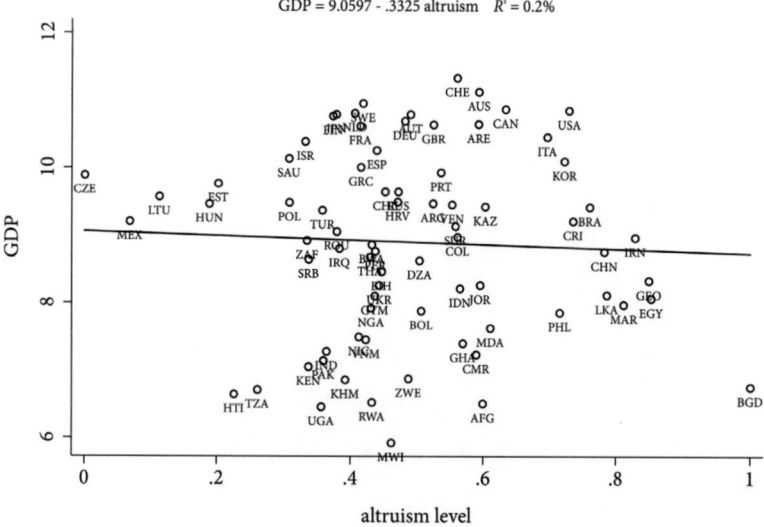

Note: log of GDP per capita in USD

Data source: Falk et al., 2018; Falk, Becker, Dohmen, Huffman, et al., 2016; World Bank, 2017

D.4 Spatial Distribution of Preferences in India

Figure D.8: Distribution of patience among Indian states

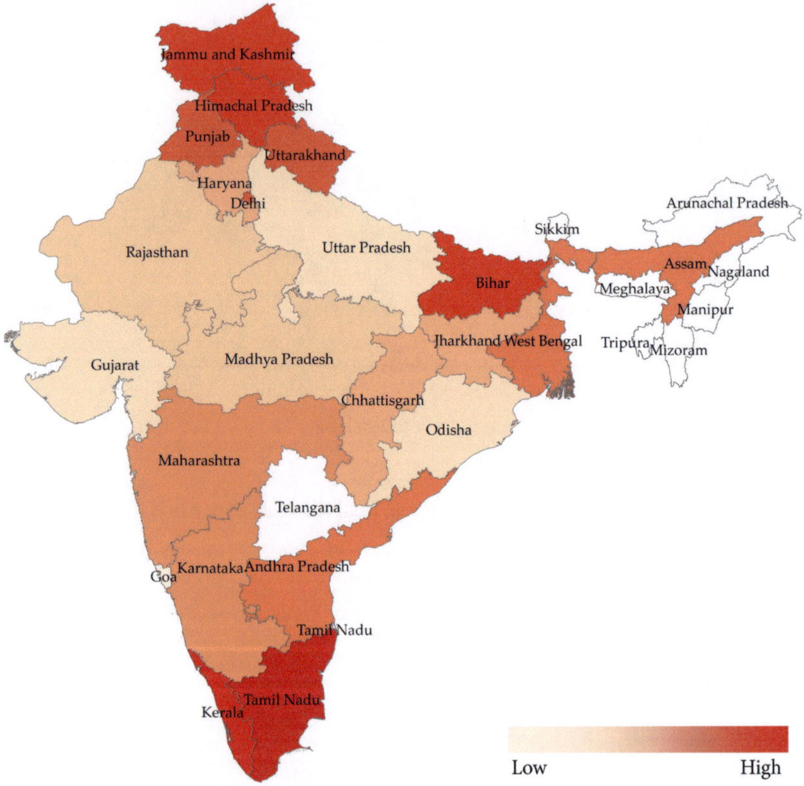

Data source: Falk et al., 2018; Falk, Becker, Dohmen, Huffman, et al., 2016. Author's illustration

Spatial Distribution of Preferences in India 209

Figure D.9: Distribution of risk taking among Indian states

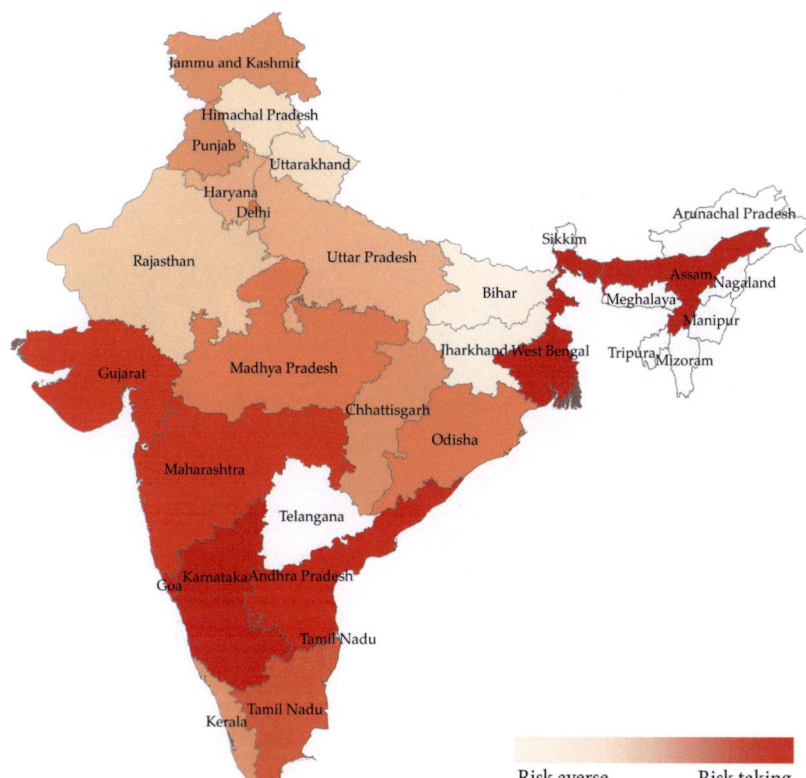

Data source: Falk et al., 2018; Falk, Becker, Dohmen, Huffman, et al., 2016. Author's illustration

Figure D.10: Distribution of altruism among Indian states

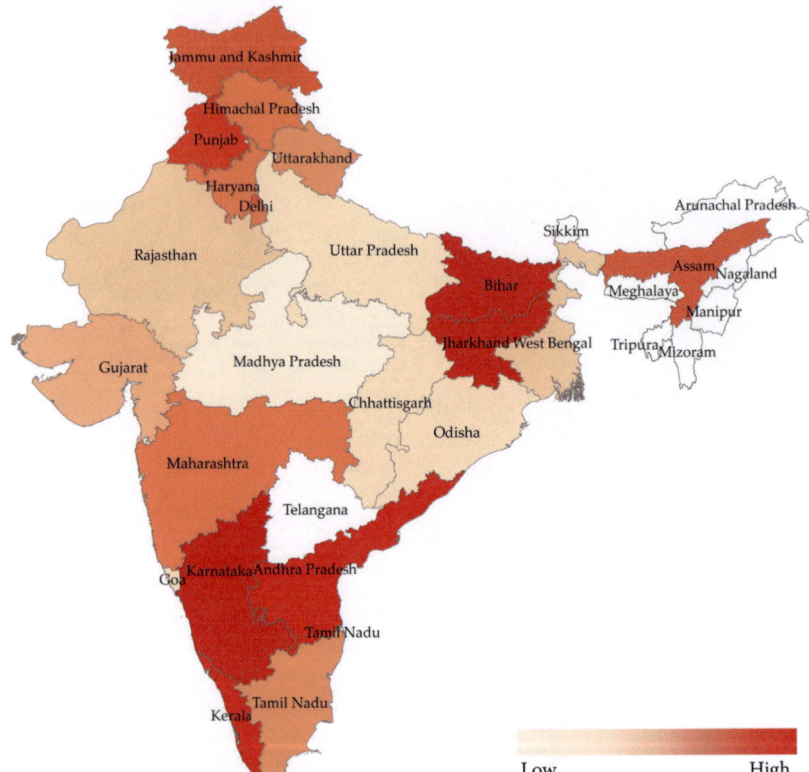

Data source: Falk et al., 2018; Falk, Becker, Dohmen, Huffman, et al., 2016. Author's illustration

D.5 Preferences and Income in India

Figure D.11: Correlation of patience and income in India

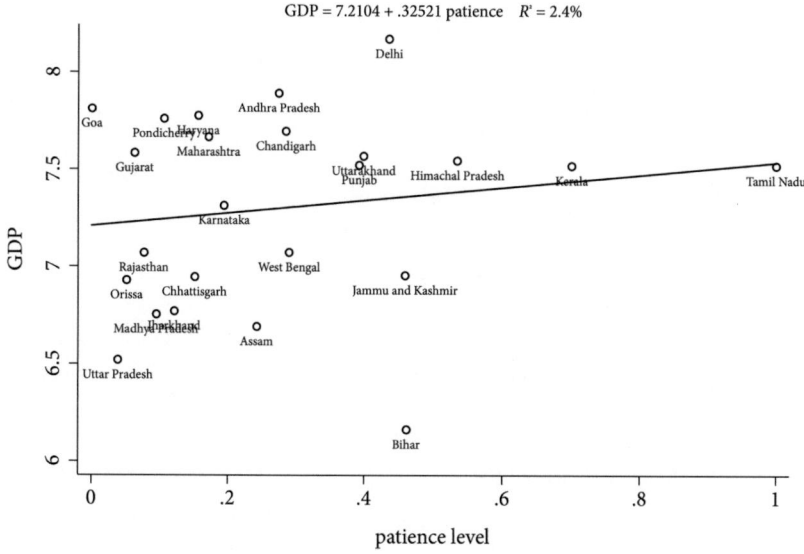

Note: log of GDP per capita in USD

Data source: Falk et al., 2018; Falk, Becker, Dohmen, Huffman, et al., 2016; Government of India, 2018b

Figure D.12: Correlation of risk and income in India

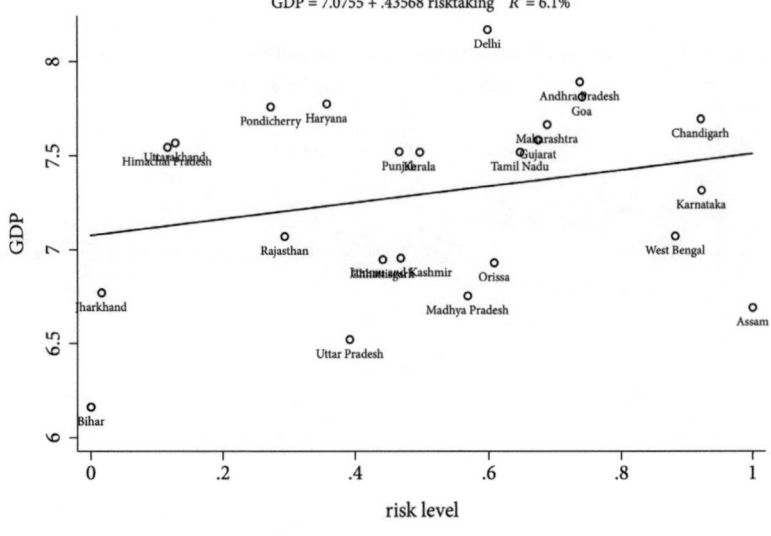

Note: log of GDP per capita in USD

Data source: Falk et al., 2018; Falk, Becker, Dohmen, Huffman, et al., 2016; Government of India, 2018b

Figure D.13: Correlation of altruism and income in India

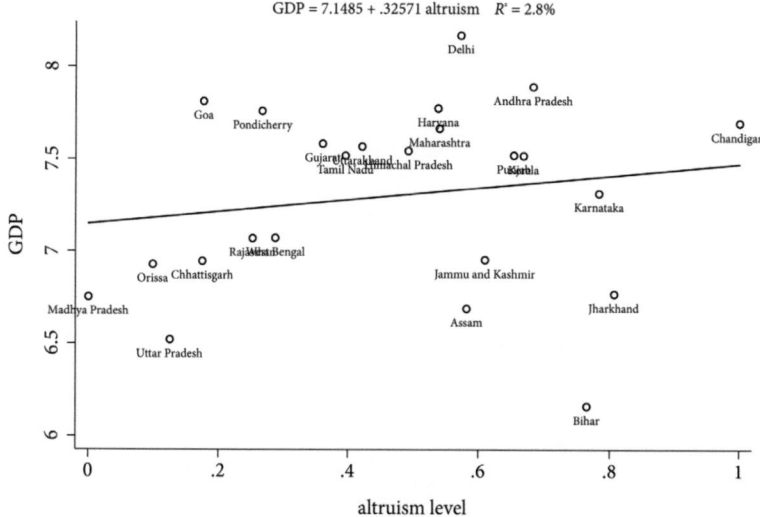

Note: log of GDP per capita in USD

Data source: Falk et al., 2018; Falk, Becker, Dohmen, Huffman, et al., 2016; Government of India, 2018b

D.6 Distribution of Preferences in India

Figure D.14: Probability distribution of patience in India

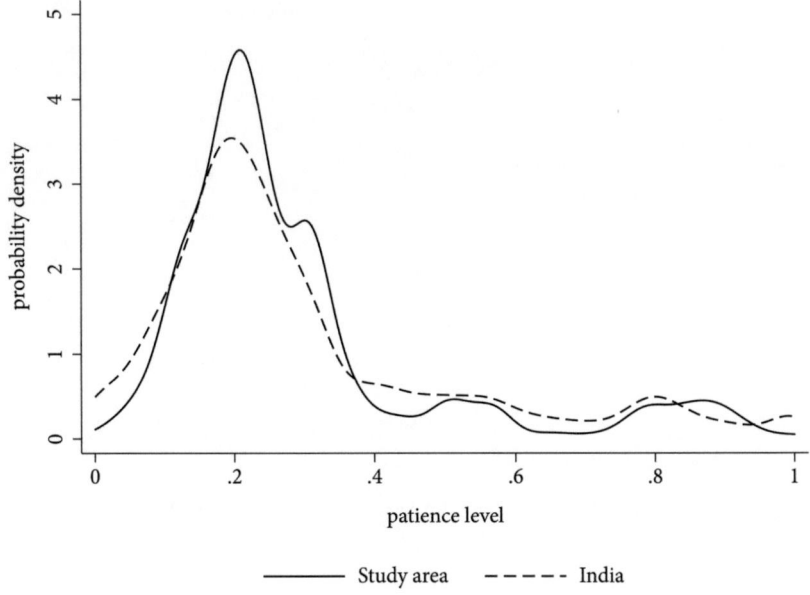

Data source: Falk et al., 2018; Falk, Becker, Dohmen, Huffman, et al., 2016. The distribution is computed as kernel density estimation using a Gaussian kernel

Distribution of Preferences in India 215

Figure D.15: Probability distribution of risk in India

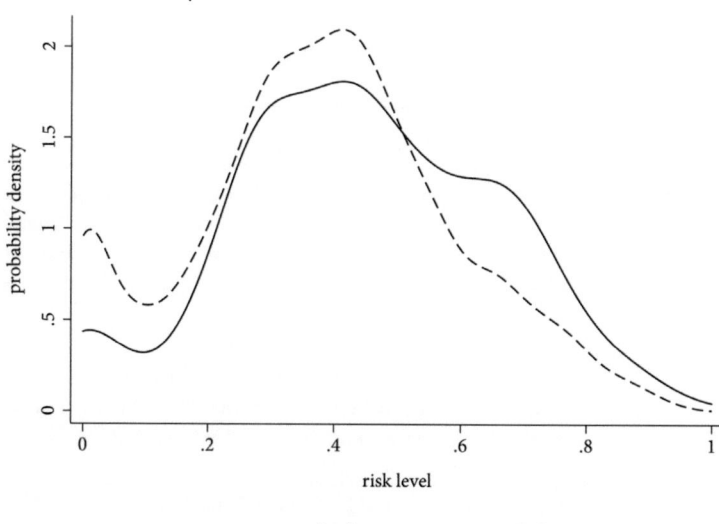

Data source: Falk et al., 2018; Falk, Becker, Dohmen, Huffman, et al., 2016. The distribution is computed as kernel density estimation using a Gaussian kernel

Figure D.16: Probability distribution of altruism in India

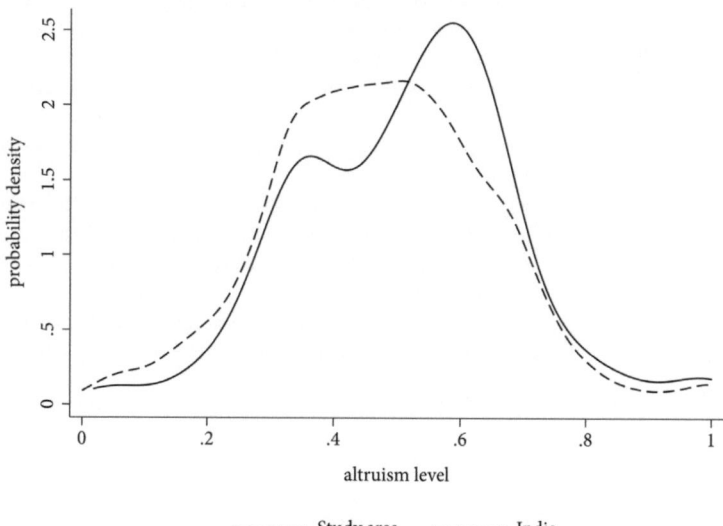

Data source: Falk et al., 2018; Falk, Becker, Dohmen, Huffman, et al., 2016. The distribution is computed as kernel density estimation using a Gaussian kernel

D.7 Survey Preference Module

Table D.1: Survey items

Preference	No.	Question
Risk	1	Staircase risk
	2	Please tell me, in general, how willing or unwilling you are to take risks, using a scale from 0 to 10, where 0 means you are "completely unwilling to take risks" and 10 means you are "very willing to take risks." You can use any number between 0 and 10 to indicate where you fall on the scale, using 0, 1, 2, 3, 4, 5, 6, 7, 8, 9, or 10.
Time	1	Staircase time
	2	How willing are you to give up something that is beneficial for you today in order to benefit more from that in the future? You can use any number between 0 and 10 to indicate where you fall on the scale, using 0, 1, 2, 3, 4, 5, 6, 7, 8, 9, or 10.
Altruism	1	How willing are you to give to good causes without expecting anything in return? You can use any number between 0 and 10 to indicate where you fall on the scale, using 0, 1, 2, 3, 4, 5, 6, 7, 8, 9, or 10.
	2	Imagine the following situation: Today you unexpectedly receive INR 2,400. How much of this amount would you donate to a good cause?
Pos. reciprocity	1	When someone does me a favor, I am willing to return it. You can use any number between 0 and 10 to indicate where you fall on the scale, using 0, 1, 2, 3, 4, 5, 6, 7, 8, 9, or 10.
	2	Please think about what you would do in the following situation. You are in an area you are not familiar with, and you realize that you lost your way. You ask a stranger for directions. The stranger offers to take you to your destination. Helping you costs the stranger about INR 40 in total. However, the stranger says he or she does not want any money from you. You have six presents with you. The cheapest present costs INR 10, the most expensive one costs INR 60. Do you give one of the presents to the stranger as a "thank you" gift? Which present do you give to the stranger? A present worth between INR 0 and INR 60.

Table D.1: Survey items

Preference	No.	Question
Neg. reciprocity	1	How willing are you to punish someone who treats you unfairly, even if there may be costs for you? You can use any number between 0 and 10 to indicate where you fall on the scale, using 0, 1, 2, 3, 4, 5, 6, 7, 8, 9, or 10.
	2	How willing are you to punish someone who treats others unfairly, even if there may be costs for you? You can use any number between 0 and 10 to indicate where you fall on the scale, using 0, 1, 2, 3, 4, 5, 6, 7, 8, 9, or 10
	3	How well does it describe you: "If I am treated very unjustly, I will take revenge at the first occasion, even if there is a cost to do so."? You can use any number between 0 and 10 to indicate where you fall on the scale, using 0, 1, 2, 3, 4, 5, 6, 7, 8, 9, or 10.

Figure D.17: Staircase risk

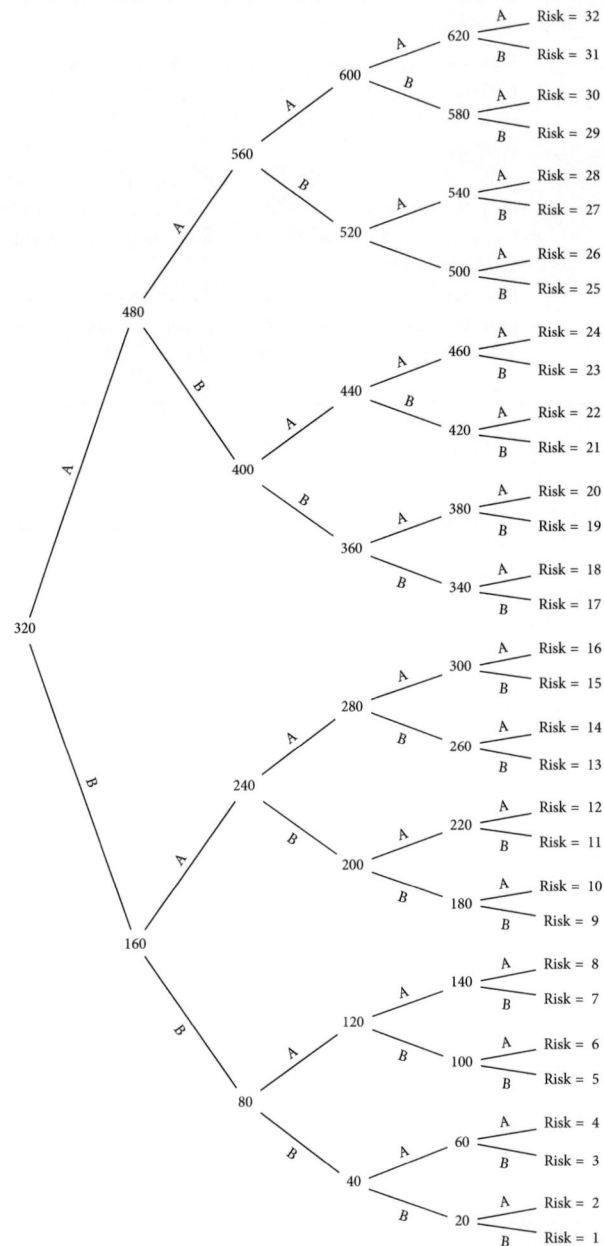

Figure D.18: Staircase time

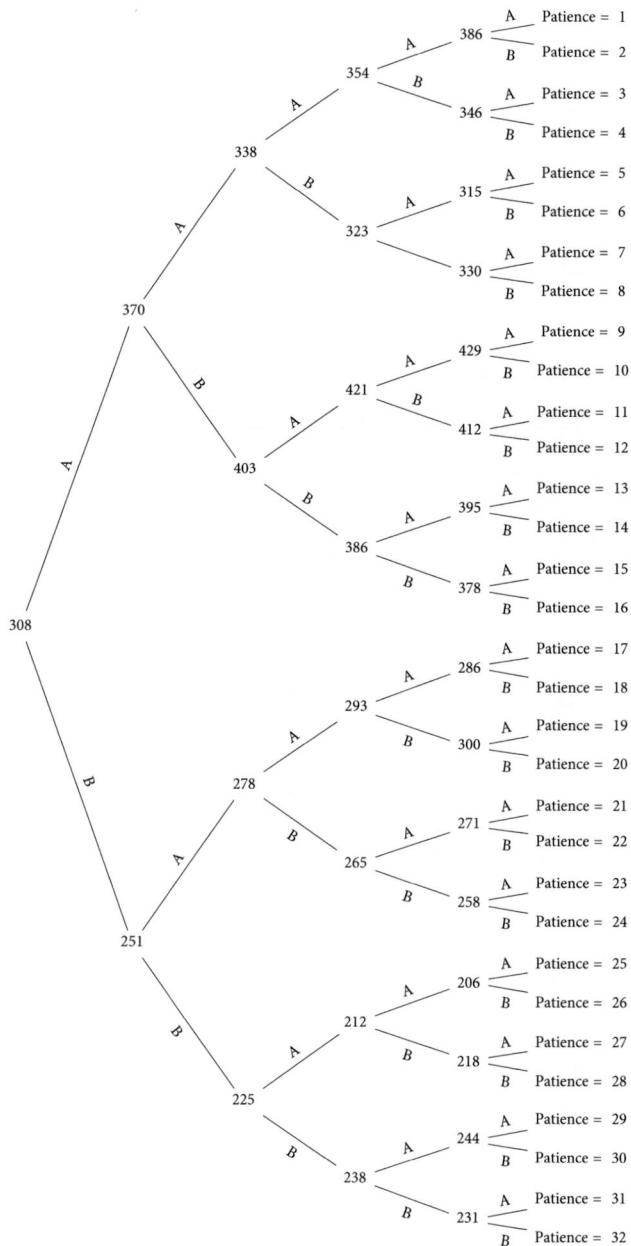

D.8 Calculation of Preferences

Risk = 0.4729985 * z-score of staircase risk

+ 0.5270015 * z-score of will. to take risks

Time = 0.7115185 * z-score of staircase time

+ 0.2884815 * z-score of will. to give something up

Altruism = 0.5350048 * z-score of will. to give to good causes

+ 0.4649952 * z-score of hypothetical donation

Pos. reciprocity = 0.4847038 * z-score of will. to to return favor

+ 0.5152962 * z-score of size of gift

Neg. reciprocity = $\dfrac{0.5261938}{2}$ * z-score of will. to punish if oneself treated unfairly

+ $\dfrac{0.5261938}{2}$ * z-score of will. to punish if other treated unfairly

+ 0.3738062 * z-score of will. to take revenge

D.9 Distribution of Preferences

Time Preference

Figure D.19: Time preference among gender

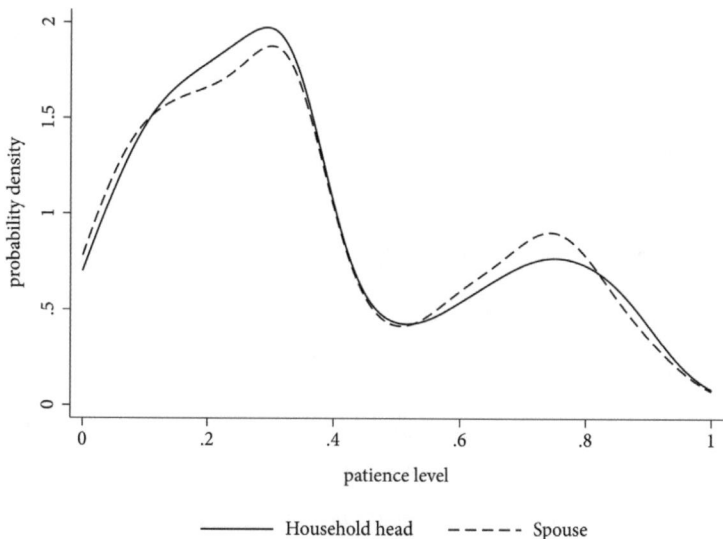

Figure D.20: Time preference among households

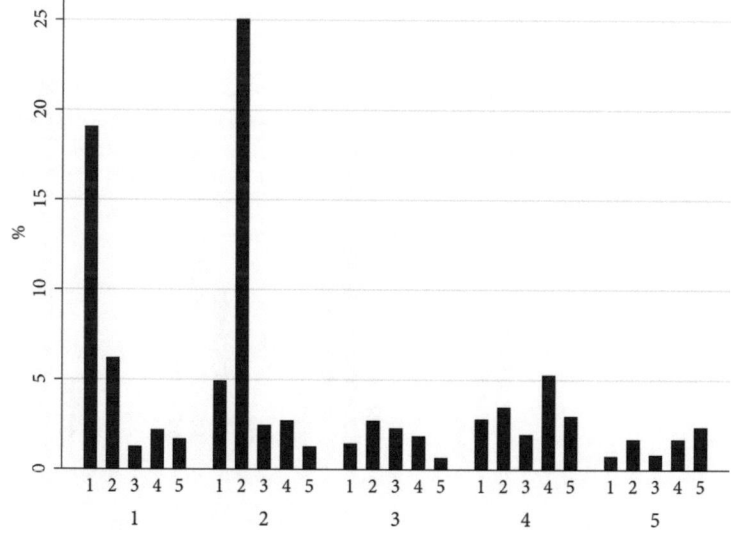

Positive Reciprocity

Figure D.21: Positive reciprocity among gender

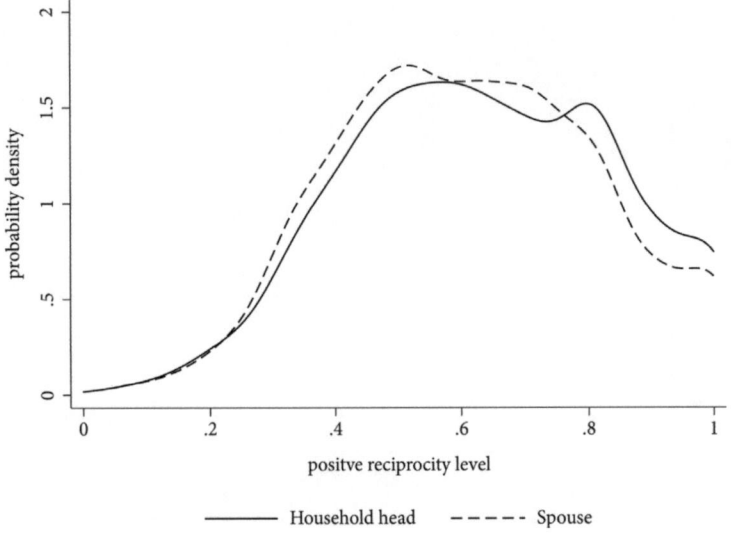

Figure D.22: Positive reciprocity among households

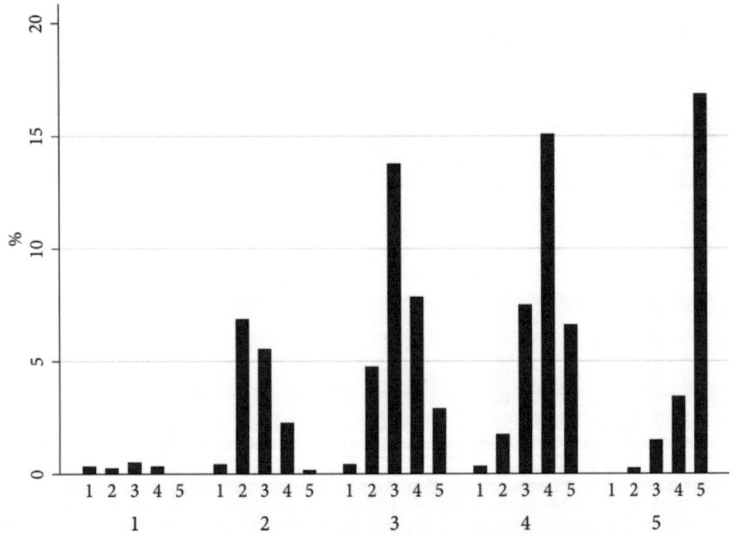

Negative Reciprocity

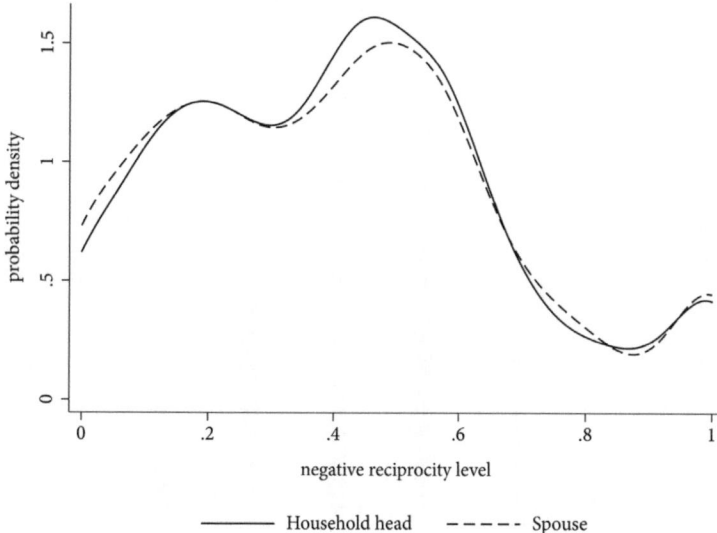

Figure D.23: Negative reciprocity among gender

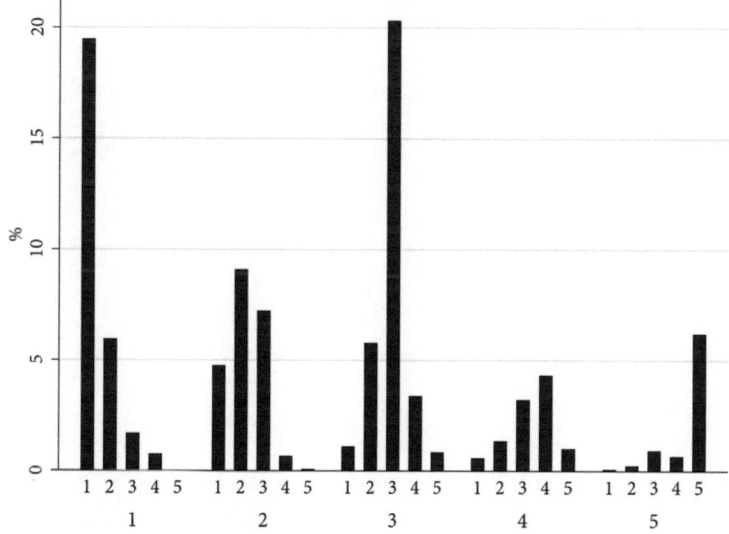

Figure D.24: Negative reciprocity among households

D.10 Summary Statistics

Figure D.25: Distribution of DDW

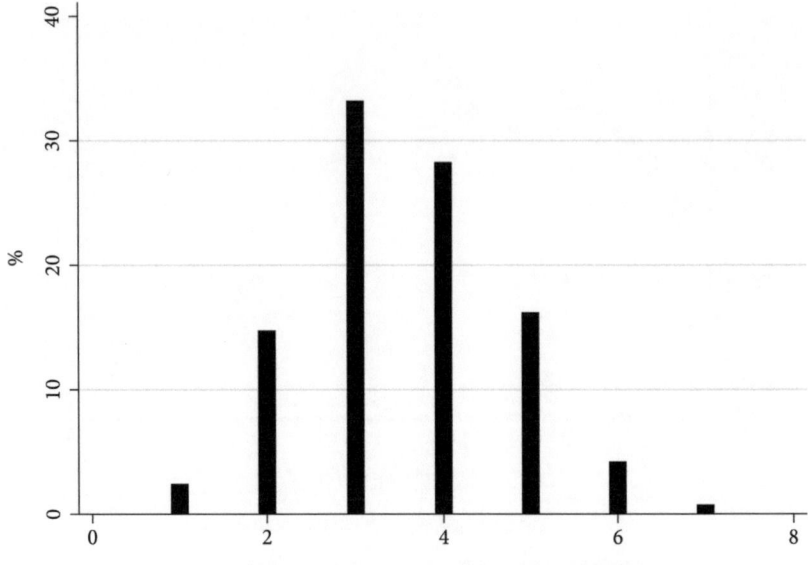

Figure D.26: Distribution of HDDS

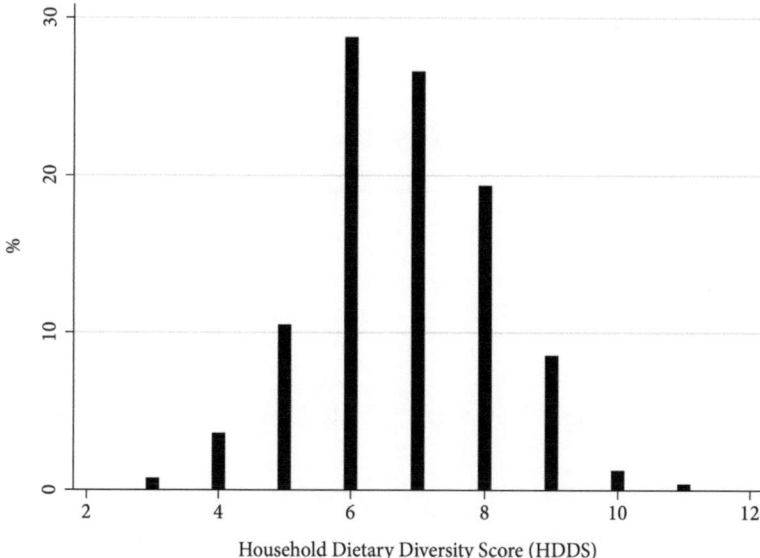

Figure D.27: Distribution of FIES

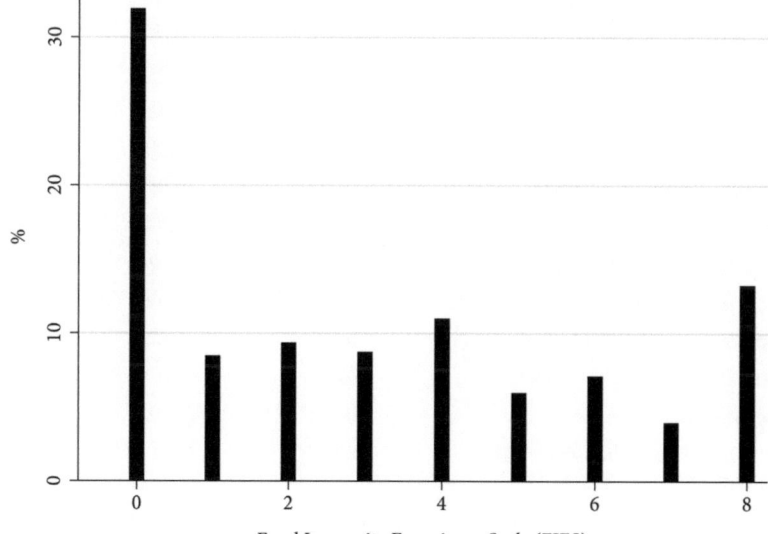

D.11 Robustness Checks

Different covariates

Table D.2: Robustness checks with different covariates (different individual variables)

Dependent variable	(1) OLS (log) DDW	
Individual preferences		
Risk level of spouse	0.087**	(0.041)
Altruism of household head	0.169***	(0.061)
Individual variables		
Age of spouse	0.001	(0.002)
Formal education of spouse	0.009***	(0.003)
Household variables		
Age of household head	0.002*	(0.001)
Literacy of household head	0.012	(0.024)
Female headed household	−0.053	(0.071)
Religion: Muslim	0.056*	(0.033)
Number of males 0-5 years	−0.033*	(0.019)
Number of males 5-15 years	0.024	(0.017)
Number of males 15-60 years	−0.017	(0.017)
Number of males 60+ years	−0.010	(0.034)
Number of females 0-5 years	−0.036*	(0.020)
Number of females 5-15 years	−0.000	(0.016)
Number of females 15-60 years	0.015	(0.018)
Number of females 60+ years	−0.017	(0.034)
(log) Income	0.070***	(0.021)
(log) Total value of liquidable assets	0.021**	(0.008)

Table D.2: *(continued)*

	(1) OLS (log) DDW	
Dependent variable		
Non-farm occupation	−0.040	(0.029)
Regular market visits	−0.024	(0.023)
Number of government schemes	0.010	(0.009)
Village variables		
(log) Village population	0.035**	(0.014)
Years that village is electrified	0.002	(0.001)
Village infrastructure	0.015	(0.012)
Day labor employment situation	0.007	(0.013)
Women group	0.001	(0.000)
NGO support	−0.026*	(0.015)
Years to last covariate shock	0.001	(0.001)
Observations	1098	
Adjusted R^2	0.193	

Robust standard errors clustered by household in parentheses

*** $p<0.01$, ** $p<0.05$, * $p<0.1$

Table D.3: Robustness checks with different covariates (different household variables)

Dependent variable	(1) OLS (log) DDW	
Individual preferences		
Risk level of spouse	0.181***	(0.050)
Altruism of household head	0.117	(0.077)
Individual variables		
Age of spouse	0.001	(0.002)
Literacy of spouse	0.101***	(0.030)
Household variables		
Age of household head	0.002	(0.001)
Formal education of household head	−0.002	(0.004)
Female headed household	−0.121	(0.097)
Religion: Muslim	0.071	(0.049)
Household members	0.000	(0.009)
(log) Income	0.079***	(0.023)
Total land size	0.005	(0.010)
Non-farm occupation	−0.029	(0.048)
Regular market visits	0.023	(0.028)
Household is electrified	0.103**	(0.045)
Number of government schemes	0.020*	(0.010)
Village variables		
(log) Village population	0.045**	(0.018)
Years that village is electrified	0.002*	(0.001)
Village infrastructure	0.023	(0.016)
Day labor employment situation	0.041**	(0.018)
Women group	0.001**	(0.001)
NGO support	−0.055***	(0.020)
Years to last covariate shock	0.000	(0.001)
Observations	724	
Adjusted R^2	0.232	

Robust standard errors clustered by household in parentheses
*** $p<0.01$, ** $p<0.05$, * $p<0.1$

Table D.4: Robustness checks with different covariates (different village variables)

Dependent variable	(1) OLS (log) DDW	
Individual preferences		
Risk level of spouse	0.081**	(0.041)
Altruism of household head	0.168***	(0.062)
Individual variables		
Age of spouse	0.000	(0.001)
Literacy of spouse	0.080***	(0.027)
Household variables		
Age of household head	0.002*	(0.001)
Literacy of household head	0.008	(0.024)
Female headed household	−0.041	(0.071)
Religion: Muslim	0.071**	(0.032)
Number of males 0-5 years	−0.038**	(0.019)
Number of males 5-15 years	0.021	(0.017)
Number of males 15-60 years	−0.022	(0.017)
Number of males 60+ years	−0.014	(0.034)
Number of females 0-5 years	−0.039**	(0.019)
Number of females 5-15 years	−0.003	(0.016)
Number of females 15-60 years	0.014	(0.017)
Number of females 60+ years	−0.019	(0.034)
(log) Income	0.058***	(0.021)
(log) Total value of liquidable assets	0.021***	(0.008)
Non-farm occupation	−0.019	(0.030)
Regular market visits	−0.021	(0.023)
Number of government schemes	0.009	(0.009)
(log) Village population 10 years ago	0.026**	(0.013)

Table D.4: *(continued)*

Dependent variable	(1) OLS (log) DDW	
Years that village is electrified	0.002^{***}	(0.001)
Village infrastructure 10 years ago	-0.002	(0.020)
Day labor employment situation 10 years ago	-0.012	(0.015)
Women group 10 years ago	-0.002^{**}	(0.001)
NGO support 10 years ago	0.040^{*}	(0.021)
Years to last covariate shock	0.002^{*}	(0.001)
Observations	1095	
Adjusted R^2	0.193	

Robust standard errors clustered by household in parentheses

*** p<0.01, ** p<0.05, * p<0.1

Table D.5: Robustness checks with different covariates (different individual, household and village variables)

Dependent variable	(1) OLS (log) DDW	
Individual preferences		
Risk level of spouse	0.204***	(0.051)
Altruism of household head	0.148*	(0.079)
Individual variables		
Age of spouse	−0.000	(0.002)
Formal education of spouse	0.005	(0.004)
Household variables		
Age of household head	0.002	(0.001)
Formal education of household head	0.001	(0.004)
Female headed household	−0.122	(0.081)
Religion: Muslim	0.136***	(0.048)
Household members	−0.004	(0.009)
(log) Income	0.068***	(0.023)
Total land size	0.010	(0.009)
Non-farm occupation	−0.004	(0.051)
Regular market visits	0.016	(0.029)
Household is electrified	0.101**	(0.044)
Number of government schemes	0.018*	(0.011)
(log) Village population 10 years ago	0.022	(0.018)
Years that village is electrified	0.004***	(0.001)
Village infrastructure 10 years ago	−0.019	(0.024)
Day labor employment situation 10 years ago	0.022	(0.024)
Women group 10 years ago	−0.002***	(0.001)
NGO support 10 years ago	0.044*	(0.023)
Years to last covariate shock	0.001	(0.001)
Observations	720	
Adjusted R^2	0.218	

Robust standard errors clustered by household in parentheses
*** $p<0.01$, ** $p<0.05$, * $p<0.1$

Checks for multicollinearity

Table D.6: VIF for multicollinearity test

	(1) OLS		
Dependent variable	(log) DDW		
	β-coefficient	SE	VIF
Individual preferences			
Risk level of spouse	0.119***	(0.043)	1.09
Altruism of household head	0.186***	(0.064)	1.19
Individual variables			
Age of spouse	0.001	(0.002)	3.39
Literacy of spouse	0.080***	(0.028)	1.60
Household variables			
Age of household head	0.002*	(0.001)	2.38
Literacy of household head	0.008	(0.025)	1.32
Female headed household	−0.039	(0.070)	1.06
Religion: Muslim	0.065*	(0.034)	2.90
Number of males 0-5 years	−0.038*	(0.020)	1.56
Number of males 5-15 years	0.024	(0.018)	1.08
Number of males 15-60 years	−0.022	(0.018)	1.51
Number of males 60+ years	−0.016	(0.037)	1.51
Number of females 0-5 years	−0.042**	(0.020)	1.67
Number of females 5-15 years	−0.010	(0.016)	1.11
Number of females 15-60 years	0.023	(0.019)	1.38
Number of females 60+ years	−0.019	(0.037)	1.27
(log) Income	0.095***	(0.022)	1.88
(log) Total value of liquidable assets	0.026***	(0.009)	2.81
Non-farm occupation	−0.050*	(0.030)	1.41
Regular market visits	−0.020	(0.024)	1.12
Number of government schemes	0.017*	(0.010)	1.26

Table D.6: *(continued)*

	(1) OLS (log) DDW		
Dependent variable			
	β-coefficient	SE	VIF
Village variables			
(log) Village population	0.035**	(0.015)	2.54
Years that village is electrified	0.002*	(0.001)	4.24
Village infrastructure	0.010	(0.012)	1.56
Day labor employment situation	0.013	(0.013)	2.20
Women group	0.001*	(0.001)	1.86
NGO support	−0.029*	(0.016)	2.94
Years to last covariate shock	0.000	(0.001)	1.86
Observations	1102		
Adjusted R^2	0.239		
Mean VIF	1.85		

Robust standard errors clustered by household in parentheses

*** $p<0.01$, ** $p<0.05$, * $p<0.1$

Bibliography

Allen, Summer and Alan de Brauw
    2018   "Nutrition sensitive value chains: Theory, progress, and open questions", *Global Food Security*, 16, pp. 22-28.

Anderson, Lisa R. and Jennifer M. Mellor
    2008   "Predicting health behaviors with an experimental measure of risk preference", *Journal of Health Economics*, 27, 5, pp. 1260-1274.

Andreoni, James
    1995   "Warm-Glow versus Cold-Prickle: The Effects of Positive and Negative Framing on Cooperation in Experiments", *The Quarterly Journal of Economics*, 110, 1, pp. 1-21.

Arimond, Mary and Marie T. Ruel
    2004   "Dietary Diversity Is Associated with Child Nutritional Status: Evidence from 11 Demographic and Health Surveys", *The Journal of nutrition*, 134, 10, pp. 2579-2585.

Arrow, Kenneth J.
    1965   *Aspects of the Theory of Risk Bearing*, Yrjö Jahnssonin Säätiö, Helsinki.

Bailey, Regan L., Keith P. West, and Robert E. Black
    2015   "The Epidemiology of Global Micronutrient Deficiencies", *Annals of Nutrition and Metabolism*, 66, suppl 2, pp. 22-33.

Ballard, Terri J., Anne W. Kepple, and Carlo Cafiero
    2013   *The Food Insecurity Experience Scale: Development of a Global Standard for Monitoring Hunger Worldwide*, FAO, Rome.

Bandiera, Oriana, Iwan Barankay, and Imran Rasul
    2005   "Social Preferences and the Response to Incentives: Evidence from Personnel Data", *The Quarterly Journal of Economics*, 120, 3, pp. 917-962.
    2009   "Social Connections and Incentives in the Workplace: Evidence From Personnel Data", *Econometrica*, 77, 4, pp. 1047-1094.

Barlow, Pepita, Aaron Reeves, Martin McKee, Gauden Galea, and David Stuckler
: 2016 "Unhealthy Diets, Obesity and Time Discounting: a Systematic Literature Review and Network Analysis", *Obesity Reviews*, 11, pp. 1-10.

Barrett, Christopher B.
: 1999 "The Microeconomics of the Developmental Paradox: on the Political Economy of Food Price Policy", *Agricultural Economics*, 20, 2, pp. 159-172.
: 2010 "Measuring Food Insecurity", *Science*, 327, 5967, pp. 825-828.

Barsky, Robert B., F. Thomas Juster, Miles S. Kimball, and Matthew D. Shapiro
: 1997 "Preference Parameters and Behavioral Heterogeneity: An Experimental Approach in the Health and Retirement Study", *The Quarterly Journal of Economics*, 112, 2, pp. 537-579.

Becker, Gary S.
: 1967 "Human Capital and the Personal Distribution of Income: an Analytical Approach", Ann Arbor.
: 1974 "A Theory of Social Interactions", *Journal of Political Economy*, 82, 6, pp. 1063-1093.
: 1981 "Altruism in the Family and Selfishness in the Market Place", *Economica*, 48, 189, pp. 1-15.

Becker, Gary S. and Robert J. Barro
: 1986 "Altruism and the Economic Theory of Fertility", *Population and Development Review*, 12, 1986, pp. 69-76.

Ben-Porath, Yoram
: 1967 "The Production of Human Capital and the Life Cycle of Earnings", *Journal of Political Economy*, 75, 4, pp. 352-365.

Bergstrom, Ted and Oded Stark
: 1993 "How Altruism Can Prevail in an Evolutionary Environment", *American Economic Review*, 83, 2, pp. 149-155.

Berti, Peter R.
: 2015 "Relationship Between Production Diversity and Dietary Diversity Depends on How Number of Foods is Counted", *Proceedings of the National Academy of Sciences*, 112, 42, E5656-E5656.

Biesalski, Hans Konrad
   2013   *Hidden Hunger*, Springer, Berlin.
   2015   *Mikronährstoffe als Motor der Evolution*, Springer, Berlin.

Binswanger, Hans P.
   1981   "Attitudes Toward Risk: Theoretical Implications of an Experiment in Rural India", *The Economic Journal*, 91, 364, p. 867.

Black, Robert E., Lindsay H. Allen, Zulfiqar A. Bhutta, Laura E. Caulfield, Mercedes de Onis, Majid Ezzati, Colin Mathers, and Juan Rivera
   2008   "Maternal and child undernutrition: global and regional exposures and health consequences", *The Lancet*, 371, 9608, pp. 243-260.

Black, Robert E., Cesar G. Victora, Susan P. Walker, Zulfiqar A. Bhutta, Parul Christian, Mercedes de Onis, Majid Ezzati, Sally Grantham-McGregor, Joanne Katz, Reynaldo Martorell, and Ricardo Uauy
   2013   "Maternal and child undernutrition and overweight in low-income and middle-income countries", *The Lancet*, 382, 9890, pp. 427-451.

Bleakley, Hoyt
   2010   "Health, Human Capital, and Development", *Annual Review of Economics*, 2, 1, pp. 283-310.

Borghans, Lex and Bart H.H. Golsteyn
   2006   "Time Discounting and the Body Mass Index: Evidence from the Netherlands", *Economics and Human Biology*, 4, 1, pp. 39-61.

Bosworth, Barry and Susan M. Collins
   2008   "Accounting for Growth: Comparing China and India", *Journal of Economic Perspectives*, 22, 1, pp. 45-66.

Bouis, Howarth E., Patrick Eozenou, and Aminur Rahman
   2011   "Food Prices, Household Income, and Resource Allocation: Socioeconomic Perspectives on Their Effects on Dietary Quality and Nutritional Status", *Food and Nutrition Bulletin*, 32, 1 (supplement), S14-S23.

Brückner, Markus and Antonio Ciccone
   2011   "Rain and the Democratic Window of Opportunity", *Econometrica*, 79, 3, pp. 923-947.

Burks, Stephen V., Jeffrey P. Carpenter, Lorenze Goette, and Aldo Rustichini
  2009 "Cognitive Skills Affect Economic Preferences, Strategic Behavior, and Job Attachment", *Proceedings of the National Academy of Sciences*, 106, 19, pp. 7745-7750.

Cafiero, Carlo, Hugo R. Melgar-Quinonez, Terri J. Ballard, and Anne W. Kepple
  2014 "Validity and Reliability of Food Security Measures", *Annals of the New York Academy of Sciences*, 1331, 1, pp. 230-248.

Castillo, Marco, Jeffrey L. Jordan, and Ragan Petrie
  2018 "Children's Rationality, Risk Attitudes and Field Behavior", *European Economic Review*, 102, pp. 62-81.

CFS
  2017 *Global Strategic Framework for Food Security and Nutrition (GSF)*, FAO, Rome.

Charness, Gary and Matthew Rabin
  2002 "Understanding Social Preferences with Simple Tests", *The Quarterly Journal of Economics*, 117, 3, pp. 817-869.

Chege, Christine G. K., Camilla I. M. Andersson, and Matin Qaim
  2015 "Impacts of Supermarkets on Farm Household Nutrition in Kenya", *World Development*, 72, pp. 394-407.

Chetty, Raj and Adam Szeidl
  2007 "Consumption Commitments and Risk Preferences", *The Quarterly Journal of Economics*, 122, 2, pp. 831-877.

Choi, Syngjoo, Shachar Kariv, Wieland Müller, and Dan Silverman
  2014 "Who Is (More) Rational?", *American Economic Review*, 104, 6, pp. 1518-1550.

Croson, Rachel and Uri Gneezy
  2009 "Gender Differences in Preferences", *Journal of Economic Literature*, 47, 2, pp. 448-474.

Deaton, Angus
  1992 *Understanding Consumption*, Oxford University Press, Oxford.
  2003 "Health, Inequality, and Economic Development", *Journal of Economic Literature*, 41, 1, pp. 113-158.

DeLong, J. Bradford
    2003   "India Since Independence: An Analytic Growth Narrative", in *In Search of Prosperity: Analytic Narratives on Economic Growth*, ed. by Dani Rodrik, Princeton University Press, Princeton, NJ.

Digman, John M.
    1990   "Personality Structure: Emergence of the Five-Factor Model", *Annual Review of Psychology*, 41, 1, pp. 417-440.

Dillon, Andrew, Kevin McGee, and Gbemisola Oseni
    2015   "Agricultural Production, Dietary Diversity and Climate Variability", *Journal of Development Studies*, 51, 8, pp. 976-995.

Dohmen, Thomas, Armin Falk, Bart H. H. Golsteyn, David Huffman, and Uwe Sunde
    2017   "Risk Attitudes Across The Life Course", *The Economic Journal*, 127, 605, F95-F116.

Dohmen, Thomas, Armin Falk, David Huffman, Uwe Sunde, Jürgen Schupp, and Gert G. Wagner
    2011   "Individual risk attitudes: Measurement, determinants, and behavioral consequences", *Journal of the European Economic Association*, 9, 3, pp. 522-550.

Doran, George T.
    1981   "There's a S.M.A.R.T. Way to Write Management's Goals and Objectives", *Management Review*, 70, 11, pp. 35-36.

Epstein, Leonhard H., Noelle Jankowiak, Henry Lin, Rocco Paluch, Mikhail N. Koffarnus, and Warren K. Bickel
    2014   "No Food for Thought: Moderating Effects of Delay Discounting and Future Time Perspective on the Relation between Income and Food Insecurity", *American Journal of Clinical Nutrition*, 100, 3, pp. 884-890.

Falk, Armin, Anke Becker, Thomas Dohmen, Benjamin Enke, David Huffman, and Uwe Sunde
    2015   *The Nature and Predictive Power of Preferences: Global Evidence*, IZA Discussion Paper No. 9504, IZA, Bonn.
    2017   *Global Evidence on Economic Preferences*, NBER Working Paper No. 23943, National Bureau of Economic Research, Cambridge, MA.

2018 "Global Evidence on Economic Preferences", *The Quarterly Journal of Economics*, 133, 4, pp. 1645-1692.

Falk, Armin, Anke Becker, Thomas Dohmen, David Huffman, and Uwe Sunde

2016 *The Preference Survey Module: A Validated Instrument for Measuring Risk, Time, and Social Preferences*, IZA Discussion Paper No. 9674, IZA, Bonn.

Falk, Armin and Johannes Hermle

2018 "Relationship of gender differences in preferences to economic development and gender equality", *Science*, 362, 6412, eaas9899.

FAO

2015 *Designing Nutrition-Sensitive Agriculture Investments*, FAO, Rome.

2016 *Pulses for food security and nutrition: How can their full potential be tapped?*, FAO, Rome.

2018a *FAO Food Price Index*, http://www.fao.org/worldfoodsituation/foodpricesindex/en/, accessed 4.11.2018.

2018b *FAOSTAT Statistics Database*, FAO, Rome.

FAO and FHI 360

2016 *Minimum Dietary Diversity for Women: A Guide for Measurement*, FAO, Rome.

FAO, IFAD, UNICEF, WFP, and WHO

2018 *The State of Food Security and Nutrition in the World 2018. Building climate resilience for food security and nutrition*, FAO, Rome.

FAO, IFAD, and WFP

2015 *The State of Food Insecurity in the World. Meeting the 2015 International Hunger Targets: Taking Stock of Uneven Progress*, FAO, Rome.

FAO and WHO

2004 *Vitamin and Mineral Requirements in Human Nutrition*, WHO, Geneva.

Fehr, Ernst and Klaus M. Schmidt

2006 "The Economics of Fairness, Reciprocity and Altruism – Experimental Evidence and New Theories", in *Handbook of the Economics of Giving, Altruism and Reciprocity*, ed. by Serge-Christophe Kolm and Jean Mercier Ythier, North Holland, Amsterdam, chap. 8, pp. 615-691.

Forsythe, Robert, Joel L. Horowitz, N.E. Savin, and Martin Sefton
 1994 "Fairness in Simple Bargaining Experiments", *Games and Economic Behavior*, 6, 3, pp. 347-369.

Foster, Andrew and Mark Rosenzweig
 2001 "Imperfect Commitment, Altruism, And The Family: Evidence From Transfer Behavior In Low-Income Rural Areas", *The Review of Economics and Statistics*, 83, 3, pp. 389-407.

Fox, John A.
 2011 "Risk Preferences and Food Consumption", in *The Oxford Handbook of the Economics of Food Consumption and Policy*, ed. by Jason L. Lusk, Jutta Roosen, and Jason F. Shogren, Oxford University Press, Oxford, chap. 3, pp. 75-98.

Franck, Caroline, Sonia M. Grandi, and Mark J. Eisenberg
 2013 "Taxing Junk Food to Counter Obesity", *American Journal of Public Health*, 103, 11, pp. 1949-1953.

Frederick, Shane, George Loewenstein, and Ted O'donoghue
 2002 "Time Discounting and Time Preference: A Critical Review", *Journal of Economic Literature*, 40, 2, pp. 351-401.

Frison, Emile A., Jeremy Cherfas, and Toby Hodgkin
 2011 "Agricultural Biodiversity is Essential for a Sustainable Improvement in Food and Nutrition Security", *Sustainability*, 3, 12, pp. 238-253.

Funk, Chris, Pete Peterson, Martin Landsfeld, Diego Pedreros, James Verdin, Shraddhanand Shukla, Gregory Husak, James Rowland, Laura Harrison, Andrew Hoell, and Joel Michaelsen
 2015 "The Climate Hazards Infrared Precipitation with Stations. A New eEvironmental Record for Monitoring Extremes", *Scientific Data*, 2, p. 150066.

Gatzweiler, Franz W. and Joachim von Braun
 2016 *Technological and Institutional Innovations for Marginalized Smallholders in Agricultural Development*, Springer, Heidelberg.

Gherardi, Laureano A. and Osvaldo E. Sala
- 2015 "Enhanced Interannual Precipitation Variability Increases Plant Functional Diversity that in turn Ameliorates Negative Impact on Productivity", *Ecology Letters*, 18, 12, pp. 1293-1300.

Gibney, Michael J., Susan A. Lanham-New, Aedin Cassidy, and Hester H. Vorster
- 2002 *Introduction to Human Nutrition*, Blackwell Science, Oxford.

Gillis, Mark T. and Paul L. Hettler
- 2007 "Hypothetical and Real Incentives in the Ultimatum Game and Andreoni's Public Goods Game: An Experimental Study", *Eastern Economic Journal*, 33, 4, pp. 491-510.

Government of India
- 2013 *Press Note on Poverty Estimates, 2011-12*, Government of India, Planning Commission, Delhi.
- 2017 *Nourishing India. National Nutrition Strategy*, tech. rep., NITI AAyog, pp. 1-112.
- 2018a *Index Numbers of Wholesale Prices in India*, http://www.eaindustry.nic.in/home.asp, accessed 12.10.2018.
- 2018b *Statistical Year Book India 2017*, Government of India, Ministry of Statistics and Programme Implementation, Delhi.

Grossman, Michael
- 1972 "On the Concept of Health Capital and the Demand for Health", *Journal of Political Economy*, 80, 2, pp. 223-255.

Gupta, Poonam, Florian Michael Blum, Dhruv Jain, Sapna Roselina John, Smriti Seth, and Amit Singhi
- 2018 *India Development Update: India's Growth Story*, World Bank Group, Washington, DC.

Hausman, Jerry A.
- 1978 "Specification Tests in Econometrics", *Econometrica*, 46, 6, pp. 1251-1271.

Hayashi, Fumio
- 2000 *Econometrics*, Princeton University Press, Princeton, NJ.

Heckman, James J.
- 2006 "Skill Formation and the Economics of Investing in Disadvantaged Children", *Science*, 312, 5782, pp. 1900-1902.

Herforth, Anna and Terri J. Ballard
- 2016 "Nutrition Indicators in Agriculture Projects: Current Measurement, Priorities, and Gaps", *Global Food Security*, 10, September 2016, pp. 1-10.

Hideki, Hashimoto
- 2013 *Stability of Preference against Aging and Health Shocks : A comparison between Japan and the United States*, RIETI Discussion Paper Series 13-E-068, RIETI, Tokyo.

Hirvonen, Kalle and John Hoddinott
- 2016 "Agricultural Production and Children's Diets: Evidence from Rural Ethiopia", *Agricultural Economics*, 48, 4, pp. 469-480.

Hirvonen, Kalle, John Hoddinott, Bart Minten, and David Stifel
- 2017 "Children's Diets, Nutrition Knowledge, and Access to Markets", *World Development*, 95, July 2017, pp. 303-315.

Hjort, Jonas
- 2014 "Ethnic Divisions and Production in Firms", *The Quarterly Journal of Economics*, 129, 4, pp. 1899-1946.

Hoddinott, John
- 2012 "Agriculture, health, and nutrition: Toward conceptualizing the linkages", in *Reshaping Agriculture for Nutrition and Health*, ed. by Shenggen Fan and Rajul Pandya-Lorch, Washington, DC, chap. 2, pp. 13-20.

Hodges, Rick J., Jean C. Buzby, and B. Bennett
- 2011 "Postharvest losses and waste in developed and less developed countries: opportunities to improve resource use", *The Journal of Agricultural Science*, 149, 51, pp. 37-45.

Holt, Charles A. and Susan K. Laury
- 2002 "Risk Aversion and Incentive Effects", *American Economic Review*, 92, 5, pp. 1644-1655.

Houthakker, Hendrik S.
- 1950 "Revealed Preference and the Utility Function", *Economica*, 17, 66, p. 159.

ICF
- 2015 *The DHS Program STATcompiler. Funded by USAID*, http://www.statcompiler.com, accessed 8.10.2018.

IFPRI
- 2015 *Global Nutrition Report 2015: Actions and Accountability to Advance Nutrition and Sustainable Development*, International Food Policy Research Institute, Washington, DC.

IIPS and ICF
- 2017 *National Family Health Survey (NFHS-4), 2015-16: India*, IIPS, Mumbai.

Jacobson, Sarah and Ragan Petrie
- 2009 "Learning from Mistakes: What Do Inconsistent Choices over Risk Tell Us?", *Journal of Risk and Uncertainty*, 38, 2, pp. 143-158.

Jaenicke, Hannah and Detlef Virchow
- 2013 "Entry Points into a Nutrition-Sensitive Agriculture", *Food Security*, 5, 5, pp. 679-692.

Jones, Andrew D.
- 2017 "On-Farm Crop Species Richness Is Associated with Household Diet Diversity and Quality in Subsistence- and Market-Oriented Farming Households in Malawi", *The Journal of Nutrition*, 147, 1, pp. 86-96.

Jones, Andrew D., Scott B. Ickes, Laura E. Smith, Mduduzi N N Mbuya, Bernard Chasekwa, Rebecca A. Heidkamp, Purnima Menon, Amanda A. Zongrone, and Rebecca J. Stoltzfus
- 2014 "World Health Organization Infant and Young Child Feeding Indicators and their Associations with Child Anthropometry: A Synthesis of Recent Findings", *Maternal and Child Nutrition*, 10, 1, pp. 1-17.

Jones, Andrew D., Aditya Shrinivas, and Rachel Bezner-Kerr
- 2014 "Farm production diversity is associated with greater household dietary diversity in Malawi: Findings from nationally representative data", *Food Policy*, 46 (June 2014), pp. 1-12.

Kahneman, Daniel, Jack L. Knetsch, and Richard H. Thaler
    1990   "Experimental Tests of the Endowment Effect and the Coase Theorem", *Journal of Political Economy*, 98, 6, pp. 1325-1348.

Kahneman, Daniel and Amos Tversky
    1979   "Prospect Theory: An Analysis of Decision under Risk", *Econometrica*, 47, 2, pp. 263-292.

Kakietek, Jakub, Julia Dayton Eberwein, Dylan Walters, and Meera Shekar
    2017   *Unleashing Gains in Economic Productivity with Investments in Nutrition*, World Bank Group, Washington, DC.

Kar, Bhoomika R., Shobini L. Rao, and B. A. Chandramouli
    2008   "Cognitive Development in Children with Chronic Protein Energy Malnutrition", *Behavioral and Brain Functions*, 4, 1, pp. 31-42.

Kavitha, Kasala, Pramanik Soumitra, and Ravula Padmaja
    2016   "Understanding the Linkages between Crop Diversity and Household Dietary Diversity in the Semi-Arid Regions of India", *Agricultural Economics Research Review*, 29, pp. 129-137.

Kearney, John
    2010   "Food Consumption Trends and Drivers", *Philosophical Transactions of the Royal Society B: Biological Sciences*, 365, 1554, pp. 2793-2807.

Keylock, Christopher J.
    2005   "Simpson Diversity and the Shannon-Wiener Index as Special Cases of a Generalized Entropy", *Oikos*, 109, 1, pp. 203-207.

Kimball, Miles S, Claudia R Sahm, and Matthew D Shapiro
    2009   "Risk Preferences in the PSID: Individual Imputations and Family Covariation", *American Economic Review*, 99, 2, pp. 363-368.

Koppmair, Stefan, Menale Kassie, and Matin Qaim
    2017   "Farm Production, Market Access and Dietary Diversity in Malawi", *Public Health Nutrition*, 20, 2, pp. 325-335.

Koppmair, Stefan and Matin Qaim
    2017a   "Farm Production Diversity and Individual-level Dietary Diversity. Response to: 'Not all dietary diversity scores can legitimately be interpreted as proxies of diet quality' by Verger et al." *Public Health Nutrition*, 20, 11, pp. 2070-2072.

2017b "Response to: 'On the appropriate use and interpretation of dietary diversity scores' by Verger et al." *Public Health Nutrition*, 20, 11, p. 2075.

Koszegi, Botond and Matthew Rabin
2006 "A Model of Reference-Dependent Preferences", *The Quarterly Journal of Economics*, 121, 4, pp. 1133-1165.

Leroy, Jef L., Marie Ruel, Edward A. Frongillo, Jody Harris, and Terri J. Ballard
2015 "Measuring the Food Access Dimension of Food Security: A Critical Review and Mapping of Indicators", *Food and Nutrition Bulletin*, 36, 2, pp. 167-195.

Levhari, David and Yoram Weiss
1974 "The Effect of Risk on the Investment in Human Capital", *The American Economic Review*, 64, 6, pp. 950-963.

Levy, Paul S. and Stanley Lemeshow
2008 *Sampling of Populations*, John Wiley & Sons, Hoboken, NJ.

Liu, Elaine M.
2013 "Time to Change What to Sow: Risk Preferences and Technology Adoption Decisions of Cotton Farmers in China", *Review of Economics and Statistics*, 95, 4 (Oct. 2013), pp. 1386-1403.

Lobb, Alexandra E., Mario Mazzocchi, and William B. Traill
2007 "Modelling risk perception and trust in food safety information within the theory of planned behaviour", *Food Quality and Preference*, 18, 2, pp. 384-395.

Lundberg, Shelly and Robert Pollak
1993 "Separate Spheres Bargaining and the Marriage Market", *Journal of Political Economy*, 101, 6, pp. 988-1010.

Lusk, Jayson L. and Keith H. Coble
2005 "Risk Perceptions, Risk Preference, and Acceptance of Risky Food", *American Journal of Agricultural Economics*, 87, 2, pp. 393-405.

Lusk, Jayson L., Jutta Roosen, and Jason F. Shogren
2011 *The Oxford Handbook of the Economics of Food Consumption and Policy*, Oxford University Press, Oxford.

Malapit, Hazel Jean L., Suneetha Kadiyala, Agnes R. Quisumbing, Kenda Cunningham, and Parul Tyagi

    2015 "Women's Empowerment Mitigates the Negative Effects of Low Production Diversity on Maternal and Child Nutrition in Nepal", *Journal of Development Studies*, 51, 8, pp. 1097-1123.

Malmendier, Ulrike and Stefan Nagel

    2011 "Depression Babies: Do Macroeconomic Experiences Affect Risk Taking?", *The Quarterly Journal of Economics*, 126, 1, pp. 373-416.

Martin-Prével, Yves, Pauline Allemand, Doris Wiesmann, Mary Arimond, Terri Ballard, Megan Deitchler, Marie-Claude Dop, Gina Kennedy, Warren T. K. Lee, and Mourad Moursi

    2015 *Moving Forward on Choosing a Standard Operational Indicator of Women's Dietary Diversity*, FAO, Rome.

Maxwell, Scott E., Ken Kelley, and Joseph R. Rausch

    2008 "Sample Size Planning for Statistical Power and Accuracy in Parameter Estimation", *Annual Review of Psychology*, 59, 1, pp. 537-563.

McCrae, Robert R. and Paul T. Costa

    1987 "Validation of the five-factor model of personality across instruments and observers." *Journal of Personality and Social Psychology*, 52, 1, pp. 81-90.

Menon, Purnima, Phuong H. Nguyen, Sneha Mani, Neha Kohli, Rasmi Avula, and Lan M. Tran

    2017 *Trends in Nutrition Outcomes, Determinants and Interventions in India (2006-2016)*, POSHAN Report No 10. IFPRI, New Delhi.

Miguel, Edward, Shanker Satyanath, and Ernest Sergenti

    2004 "Economic Shocks and Civil Conflict: An Instrumental Variables Approach", *Journal of Political Economy*, 112, 4, pp. 725-753.

Pangaribowo, Evita Hanie, Nicolas Gerber, and Maximo Torero

    2013 *Food and Nutrition Security Indicators: A Review*, ZEF Working Papers 108, Center for Development Research, University of Bonn, Bonn.

Pasricha, Sant-Rayn and Beverley-Ann Biggs

    2010 "Undernutrition among Children in South and South-East Asia", *Journal of Paediatrics and Child Health*, 46, 9, pp. 497-503.

Pinstrup-Andersen, Per
   2013   "Nutrition-Sensitive Food Systems: From Rhetoric to Action", *The Lancet*, 382, 9890, pp. 375-376.

Pratt, John W.
   1964   "Risk Aversion in the Small and in the Large", *Econometrica*, 32, 1, pp. 122-136.

Reardon, Thomas, C. Peter Timmer, and Bart Minten
   2012   "Supermarket Revolution in Asia and Emerging Development Strategies to Include Small Farmers", *Proceedings of the National Academy of Sciences*, 109, 31, pp. 12332-12337.

Remans, Roseline, Fabrice A. J. DeClerck, Gina Kennedy, and Jessica Fanzo
   2015   "Expanding the View on the Production and Dietary Diversity Link: Scale, Function, and Change over Time", *Proceedings of the National Academy of Sciences*, 112, 45, E6082-E6082.

Rieger, Matthias
   2015   "Risk aversion, time preference and health production: Theory and empirical evidence from Cambodia", *Economics & Human Biology*, 17, April 2015, pp. 1-15.

Rothman, Kenneth J.
   2012   *Epidemiology. An introduction*, 3rd Editio, Oxford University Press, Oxford.

Rotter, Julian B.
   1966   "Generalized expectancies for internal versus external control of reinforcement." *Psychological Monographs: General and Applied*, 80, 1, pp. 1-28.

Ruel, Marie T. and Harold Alderman
   2013   "Nutrition-Sensitive Interventions and Programmes: How can they Help to Accelerate Progress in Improving Maternal and Child Nutrition?", *The Lancet*, 382, 9891, pp. 536-551.

Samuelson, Paul A.
   1938   "A Note on the Pure Theory of Consumer's Behaviour", *Economica*, 5, 17, pp. 61-71.

Sanderson, Eleanor and Frank Windmeijer

    2016   "A Weak Instrument F-test in Linear IV Models with Multiple Endogenous Variables", *Journal of Econometrics*, 190, 2, pp. 212-221.

Sarsons, Heather

    2015   "Rainfall and conflict: A cautionary tale", *Journal of Development Economics*, 115, pp. 62-72.

Sen, Amartya

    1981   *Poverty and Famines: An Essay on Entitlement and Deprivation*, Clarendon Press, Oxford.

Shimadzu, Hideyasu, Maria Dornelas, Peter A. Henderson, and Anne E. Magurran

    2013   "Diversity is Maintained by Seasonal Variation in Species Abundance", *BMC Biology*, 11, 1, pp. 98-106.

Sibhatu, Kibrom T., Vijesh V. Krishna, and Matin Qaim

    2015a   "Production Diversity and Dietary Diversity in Smallholder Farm Households", *Proceedings of the National Academy of Sciences*, 112, 34, pp. 10657-10662.

    2015b   "Reply to Berti: Relationship between production and consumption diversity remains small also with modified diversity measures", *Proceedings of the National Academy of Sciences*, 112, 42, E5657-E5657.

    2015c   "Reply to Remans et al.: Strengthening markets is key to promote sustainable agricultural and food systems", *Proceedings of the National Academy of Sciences*, 112, 45, E6083-E6083.

Sibhatu, Kibrom T. and Matin Qaim

    2018a   "Farm production diversity and dietary quality: linkages and measurement issues", *Food Security*, 10, 1, pp. 47-59.

    2018b   "Review: The association between production diversity, diets, and nutrition in smallholder farm households", *Food Policy*, 77, pp. 1-18.

Singh, Inderjit, Lyn Squire, and John Strauss

    1986   "The Basic Model: Theory, Empirical Results, and Policy Conclusions", in *Agricultural Household Models*, The Johns Hopkins University Press, Baltimore, pp. 17-47.

Smith, Patricia K., Barry Bogin, and David Bishai
  2005   "Are time preference and body mass index associated? Evidence from the National Longitudinal Survey of Youth", *Economics and Human Biology*, 3, 2, pp. 259-270.

Staiger, Douglas and James H. Stock
  1997   "Instrumental Variables Regression with Weak Instruments", *Econometrica*, 65, 3, pp. 557-586.

Stark, Oded
  1995   *Altruism and Beyond*, Cambridge University Press, Cambridge.

Stigler, George J. and Gary S. Becker
  1977   "De Gustibus Non Est Disputandum", *American Economic Review*, 67, 2, pp. 76-90.

Swindale, Anne and Paula Bilinsky
  2006   *Household Dietary Diversity Score (HDDS) for Measurement of Household Food Access: Indicator Guide (Version 2)*, FANTA/FHI 360, Washington, DC.

Tanaka, Camerer and Nguyen
  2010   "Risk and Time Preferenes:Linking Experimental and Household Survey Data frome Vietnam", *American Economic Review*, 100, 1, pp. 557-571.

Taylor, J. Edward and Irma Adelman
  2003   "Agricultural Household Models: Genesis, Evolution, and Extensions", *Review of Economics of the Household*, 1, 1, pp. 33-58.

Thaler, Richard H.
  1986   "The Psychology and Economics Conference Handbook: Comments on Simon, on Einhorn and Hogarth, and on Tversky and Kahneman", *The Journal of Business*, 59, 4, S279-S284.

Thaler, Richard H. and Cass R. Sunstein
  2008   *Nudge: Improving Decisions about Health, Wealth, and Happiness*, Yale University Press, Yale.

Turner, Rachel, Corinna Hawkes, Jeff Waage, Elaine Ferguson, Farhana Haseen, Hilary Homans, Julia Hussein, Deborah Johnston, Debbi Marais, Geraldine McNeill, and Bhavani Shankar

    2013 "Agriculture for Improved Nutrition: The Current Research Landscape", *Food and Nutrition Bulletin*, 34, 4, pp. 369-377.

UNICEF

    1990 *Strategy for Improved Nutrition of Children and Women in Developing Countries*. UNICEF Policy Review, UNICEF, New York, pp. 1-38.

    2015 *UNICEF's Approach to Scaling Up Nutrition for Mothers and their Children*, Discussion paper. Programme Division, UNICEF, New York.

United Nations

    2008 *Designing Household Survey Samples: Practical Guidelines*, United Nations, Department of Economic and Social Affairs, New York.

    2017 *World Population Prospects: The 2017 Revision. Key Findings and Advance Tables*, United Nations, Department of Economic and Social Affairs, New York.

Verger, Eric O., Andrew D. Jones, Marie-Claude Dop, and Yves Martin-Prével

    2017 "On the Appropriate Use and Interpretation of Dietary Diversity Scores. Response to: 'Farm production diversity and individual-level dietary diversity' by Koppmair and Qaim", *Public Health Nutrition*, 20, 11, pp. 2067-2068.

Vermeir, Iris and Wim Verbeke

    2006 "Sustainable Food Consumption: Exploring the Consumer "Attitude – Behavioral Intention" Gap", *Journal of Agricultural and Environmental Ethics*, 19, 2, pp. 169-194.

Von Braun, Joachim

    2013 "Welternährung im Globalen Wandel", in *Rolle der Wissenschaft im Globalen Wandel. Vorträge anlässlich der Jahresversammlung vom 22. bis 24. September 2012 in Berlin*, ed. by Detlev Drenckhahn and Jörg Hacker, Wissenschaftliche Verlagsgesellschaft, Stuttgart, pp. 139-157.

    2015a "Food and Nutrition Security The Concept and its Realization", in *Bread and Brain, Education and Poverty*, ed. by A. M. Battro, I. Potrykus, and M. Sanchez Sorondo, Pontifical Academy of Sciences, Scripta Varia 125, Vatican City, pp. 69-85.

2015b *Welternährung und Nachhaltigkeit*, Oekom Verlag, Wiesbaden.

Von Braun, Joachim, Tesfaye Teklu, and Patrick Webb

1999 *Famine in Africa: Causes, Responses, and Prevention*, Published, Johns Hopkins University Press, Baltimore.

Von Weizsäcker, Carl Christian

2011 "Homo Oeconomicus Adaptivus - Die Logik des Handelns bei veränderlichen Präferenzen", in *Wohin steuert die ökonomische Wissenschaft? Ein Methodenstreit in der Volkswirtschaftslehre*, ed. by Volker Caspari and Bertram Schefold, Campus Verlag, Frankfurt/New York, pp. 221-255.

2015 "Adaptive Präferenzen und die Legitimierung dezentraler Entscheidungsstrukturen", in *Behavioral Economics und Wirtschaftspolitik, Schriften zu Ordnungsfragen der Wirtschaft. Band 100*, ed. by Christian Müller and Nils Otter, Lucius & Lucius, Stuttgart, pp. 67-98.

Wooldridge, Jeffrey M.

2013 *Introductory Econometrics: A Modern Approach*, 5th Ed. Cengage Learning, Cincinnati, OH.

World Bank

2013 *World Development Indicators 2013*, World Bank, Washington DC.

2017 *World Development Indicators 2017*, World Bank, Washington DC.

Wu, Guoyao, Beth Imhoff-Kunsch, and Amy Webb Girard

2012 "Biological Mechanisms for Nutritional Regulation of Maternal Health and Fetal Development", *Paediatric and Perinatal Epidemiology*, 26, pp. 4-26.

Yan, Hong, Cunzhu Liang, Zhiyong Li, Zhongling Liu, Bailing Miao, Chunguang He, and Lianxi Sheng

2015 "Impact of Precipitation Patterns on Biomass and Species Richness of Annuals in a Dry Steppe", *PLOS ONE*, 10, 4, e0125300.

Development Economics and Policy

Series edited by Franz Heidhues†, Joachim von Braun,
Ulrike Grote and Manfred Zeller

Vol. 1 Andrea Fadani: Agricultural Price Policy and Export and Food Production in Cameroon. A Farming Systems Analysis of Pricing Policies. The Case of Coffee-Based Farming Systems. 1999.

Vol. 2 Heike Michelsen: Auswirkungen der Währungsunion auf den Strukturanpassungsprozeß der Länder der afrikanischen Franc-Zone. 1995.

Vol. 3 Stephan Bea: Direktinvestitionen in Entwicklungsländern. Auswirkungen von Stabilisierungsmaßnahmen und Strukturreformen in Mexiko. 1995.

Vol. 4 Franz Heidhues / François Kamajou: Agricultural Policy Analysis – Proceedings of an International Seminar, held at the University of Dschang, Cameroon on May 26 and 27 1994, funded by the European Union under the Science and Technology Program (STD). 1996.

Vol. 5 Elke M. Förster: Protection or Liberalization? A Policy Analysis of the Korean Beef Sector. 1996.

Vol. 6 Gertrud Schrieder: The Role of Rural Finance for Food Security of the Poor in Cameroon. 1996.

Vol. 7 Nestor R. Ahoyo Adjovi: Economie des Systèmes de Production intégrant la Culture de Riz au Sud du Bénin: Potentialités, Contraintes et Perspectives. 1996.

Vol. 8 Jenny Müller: Income Distribution in the Agricultural Sector of Thailand. Empirical Analysis and Policy Options. 1996.

Vol. 9 Michael Brüntrup: Agricultural Price Policy and its Impact on Production, Income, Employment and the Adoption of Innovations. A Farming Systems Based Analysis of Cotton Policy in Northern Benin. 1997.

Vol. 10 Justin Bomda: Déterminants de l'Epargne et du Crédit, et leurs Implications pour le Développement du Système Financier Rural au Cameroun. 1998.

Vol. 11 John M. Msuya: Nutrition Improvement Projects in Tanzania: Implementation, Determinants of Performance, and Policy Implications. 1998.

Vol. 12 Andreas Neef: Auswirkungen von Bodenrechtswandel auf Ressourcennutzung und wirtschaftliches Verhalten von Kleinbauern in Niger und Benin. 1999.

Vol. 13 Susanna Wolf (ed.): The Future of EU-ACP Relations. 1999.

Vol. 14 Franz Heidhues / Gertrud Schrieder (eds.): Romania – Rural Finance in Transition Economies. 2000.

Vol. 15 Katinka Weinberger: Women's Participation. An Economic Analysis in Rural Chad and Pakistan. 2000.

Vol. 16 Christof Batzlen: Migration and Economic Development. Remittances and Investments in South Asia: A Case Study of Pakistan. 2000.

Vol. 17 Matin Qaim: Potential Impacts of Crop Biotechnology in Developing Countries. 2000.

Vol. 18 Jean Senahoun: Programmes d'ajustement structurel, sécurité alimentaire et durabilité agricole. Une approche d'analyse intégrée, appliquée au Bénin. 2001.

Vol. 19 Torsten Feldbrügge: Economics of Emergency Relief Management in Developing Countries. With Case Studies on Food Relief in Angola and Mozambique. 2001.

Vol. 20 Claudia Ringler: Optimal Allocation and Use of Water Resources in the Mekong River Basin: Multi-Country and Intersectoral Analyses. 2001.

Vol. 21 Arnim Kuhn: Handelskosten und regionale (Des-)Integration. Russlands Agrarmärkte in der Transformation. 2001.

Vol. 22 Ortrun Anne Gronski: Stock Markets and Economic Growth. Evidence from South Africa. 2001.

Vol. 23 Patrick Webb / Katinka Weinberger (eds.): Women Farmers. Enhancing Rights, Recognition and Productivity. 2001.

Vol. 24 Mingzhi Sheng: Lebensmittelkonsum und -konsumtrends in China. Eine empirische Analy- se auf der Basis ökonometrischer Nachfragemodelle. 2002.

Vol.	25	Maria Iskandarani: Economics of Household Water Security in Jordan. 2002.
Vol.	26	Romeo Bertolini: Telecommunication Services in Sub-Saharan Africa. An Analysis of Access and Use in the Southern Volta Region in Ghana. 2002.
Vol.	27	Dietrich Müller-Falcke: Use and Impact of Information and Communication Technologies in Developing Countries' Small Businesses. Evidence from Indian Small Scale Industry. 2002.
Vol.	28	Wolfram Erhardt: Financial Markets for Small Enterprises in Urban and Rural Northern Thailand. Empirical Analysis on the Demand for and Supply of Financial Services, with Particular Emphasis on the Determinants of Credit Access and Borrower Transaction Costs. 2002.
Vol.	29	Wensheng Wang: The Impact of Information and Communication Technologies on Farm Households in China. 2002.
Vol.	30	Shyamal K. Chowdhury: Institutional and Welfare Aspects of the Provision and Use of In-formation and Communication Technologies in the Rural Areas of Bangladesh and Peru. 2002.
Vol.	31	Annette Luibrand: Transition in Vietnam. Impact of the Rural Reform Process on an Ethnic Minority. 2002.
Vol.	32	Felix Ankomah Asante: Economic Analysis of Decentralisation in Rural Ghana. 2003.
Vol.	33	Chodechai Suwanaporn: Determinants of Bank Lending in Thailand: An Empirical Exami-nation for the Years 1992 to 1996. 2003.
Vol.	34	Abay Asfaw: Costs of Illness, Demand for Medical Care, and the Prospect of Community Health Insurance Schemes in the Rural Areas of Ethiopia. 2003.
Vol.	35	Gi-Soon Song: The Impact of Information and Communication Technologies (ICTs) on Rural Households. A Holistic Approach Applied to the Case of Lao People's Democratic Re- public. 2003.
Vol.	36	Daniela Lohlein: An Economic Analysis of Public Good Provision in Rural Russia. The Case of Education and Health Care. 2003.
Vol.	37	Johannes Woelcke. Bio-Economics of Sustainable Land Management in Uganda. 2003.

Vol. 38 Susanne M. Ziemek: The Economics of Volunteer Labor Supply. An Application to Countries of a Different Development Level. 2003.

Vol. 39 Doris Wiesmann: An International Nutrition Index. Concept and Analyses of Food Insecuri-ty and Undernutrition at Country Levels. 2004.

Vol. 40 Isaac Osei-Akoto: The Economics of Rural Health Insurance. The Effects of Formal and Informal Risk-Sharing Schemes in Ghana. 2004.

Vol. 41 Yuansheng Jiang: Health Insurance Demand and Health Risk Management in Rural China. 2004.

Vol. 42 Roukayatou Zimmermann: Biotechnology and Value-added Traits in Food Crops: Rele- vance for Developing Countries and Economic Analyses. 2004.

Vol. 43 F. Markus Kaiser: Incentives in Community-based Health Insurance Schemes. 2004.

Vol. 44 Thomas Herzfeld: *Corruption begets Corruption.* Zur Dynamik und Persistenz der Korruption. 2004.

Vol. 45 Edilegnaw Wale Zegeye: The Economics of On-Farm Conservation of Crop Diversity in Ethiopia: Incentives, Attribute Preferences and Opportunity Costs of Maintaining Local Varieties of Crops. 2004.

Vol. 46 Adama Konseiga: Regional Integration Beyond the Traditional Trade Benefits: Labor Mobility contribution. The Case of Burkina Faso and Côte d'Ivoire. 2005.

Vol. 47 Beyene Tadesse Ferenji: The Impact of Policy Reform and Institutional Transformation on Agricultural Performance. An Economic Study of Ethiopian Agriculture. 2005.

Vol. 48 Sabine Daude: Agricultural Trade Liberalization in the WTO and Its Poverty Implications. A Study of Rural Households in Northern Vietnam. 2005.

Vol. 49 Kadir Osman Gyasi: Determinants of Success of Collective Action on Local Commons. An Empirical Analysis of Community-Based Irrigation Management in Northern Ghana. 2005.

Vol.	50	Borbala E. Balint: Determinants of Commercial Orientation and Sustainability of Agricultural Production of the Individual Farms in Romania. 2006.
Vol.	51	Pamela Marinda: Effects of Gender Inequality in Resource Ownership and Access on Household Welfare and Food Security in Kenya. A Case Study of West Pokot District. 2006.
Vol.	52	Charles Palmer: The Outcomes and their Determinants from Community-Company Contracting over Forest Use in Post-Decentralization Indonesia. 2006.
Vol.	53	Hardwick Tchale: Agricultural Policy and Soil Fertility Management in the Maize-based Smallholder Farming System in Malawi. 2006.
Vol.	54	John Kedi Mduma: Rural Off-Farm Employment and its Effects on Adoption of Labor Intensive Soil Conserving Measures in Tanzania. 2006.
Vol.	55	Mareike Meyn: The Impact of EU Free Trade Agreements on Economic Development and Regional Integration in Southern Africa. The Example of EU-SACU Trade Relations. 2006.
Vol.	56	Clemens Breisinger: Modelling Infrastructure Investments, Growth and Poverty Impact. A Two-Region Computable General Equilibrium Perspective on Vietnam. 2006.
Vol.	57	Meike Wollni: Coping with the Coffee Crisis. An Analysis of the Production and Marketing Performance of Coffee Farmers in Costa Rica. 2007.
Vol.	58	Franklin Simtowe: Performance and Impact of Microfinance. Evidence from Joint Liability Lending Programs in Malawi. 2008.
Vol.	59	Xiangping Jia: Credit Rationing and Institutional Constraint. Evidence from Rural China. 2008.
Vol.	60	Holger Seebens: The Economics of Gender and the Household in Developing Countries. 2009.
Vol.	61	Stephan Piotrowski: Land Property Rights and Natural Resource Use. An Analysis of Household Behavior in Rural China. 2009.
Vol.	62	Sebastian M. Scholz: Rural Development through Carbon Finance. Forestry Projects under the Clean Development Mechanism of the

Kyoto Protocol. Assessing Smallholder Participation by Structural Equation Modeling. 2009.

Vol. 63 Jakob Rupert Friederichsen: Opening Up Knowledge Production through Participatory Research? Agricultural Research for Vietnam's Northern Uplands. 2009.

Vol. 64 Olivier Ecker: Economics of Micronutrient Malnutrition. The Demand for Nutrients in Sub-Saharan Africa. 2009.

Vol. 65 Julia Johannsen: Operational Assessment of Monetary Poverty by Proxy Means Tests. 2009

Vol. 66 Ephraim Nkonya / Nicolas Gerber / Philipp Baumgartner / Joachim von Braun / Alessandro De Pinto / Valerie Graw / Edward Kato / Julia Kloos / Teresa Walter: The Economics of Land Degradation. Toward an Integrated Global Assessment. 2011.

Vol. 67 S. Idriss Nazaire Houssou: Operational Poverty Targeting by Proxy Means Tests. Models and Policy Simulations for Malawi. 2013.

Vol. 68 Abdul Salam Lodhi: Education, Child Labor and Human Capital Formation in Selected Urban and Rural Settings of Pakistan. 2013.

Vol. 69 Evita Hanie Pangaribowo: Household Food Consumption, Women's Asset and Food Policy in Indonesia. 2013.

Vol. 70 Dan Liu: China's New Rural Cooperative Medical Scheme. Evolution, Design and Impacts. 2013.

Vol. 71 Camille Saint-Macary: Microeconomic Impacts of Institutional Change in Vietnam's Northern Uplands. Empirical Studies on Social Capital, Land and Credit Institutions. 2014.

Vol. 72 Beatrice Wambui Muriithi: Commercialization of Smallholder Horticultural Farming in Kenya. Poverty, Gender, and Institutional Arrangements. 2014.

Vol. 73 Christian C. W. Grovermann: Assessment of Pesticide Use Reduction Strategies for Thai Highland Agriculture. Combining Econometrics and Agent-based Modelling. 2015.

Vol. 74 Dawit Diriba Guta: Bio-Based Energy, Rural Livelihoods and Energy Security in Ethiopia. 2015.

Vol. 75　Tigabu Degu Getahun: Industrial Clustering, Firm Performance and Employee Welfare. Evidence from the Shoe and Flower Cluster in Ethiopia. 2016.

Vol. 76　Abu Hayat Md. Saiful Islam: Impact of Technological Innovation on the Poor. Integrated Aquaculture-Agriculture in Bangladesh. 2016.

Vol. 77　Christine Husmann: The Private Sector and the Marginalized Poor. An Assessment of the Potential Role of Business in Reducing Poverty and Marginality in Rural Ethiopia. 2016.

Vol. 78　Qui Chen: Biomass Energy Economics and Rural Livelihood in Sichuan, China. 2018.

Vol. 79　Franziska Schünemann: Economy-Wide Policy Modeling of the Food-Energy-Water Nexus. Identifying Synergies and Tradeoffs on Food, Energy, and Water Security in Malawi. 2018.

Vol. 80　Till Ludwig: Consumption Choices. The effects of food production, markets and preferences on diets in India. 2019.

www.peterlang.com